T0223778

# The Integrated Test Analysis Process for Structural Dynamic Systems

# Synthesis SEM Lectures on Experimental Mechanics

Editor
**Kristin Zimmerman,** *SEM*

Synthesis SEM Lectures on Experimental Mechanics follow the technical divisions and the direction of The Society for Experimental Mechanics (SEM). The SEM is composed of international members of academia, government, and industry who are committed to interdisciplinary application, research and development, education and active promotion of experimental methods to: (a) increase the knowledge of physical phenomena; (b) further the understanding of the behavior of materials, structures and systems; and (c) provide the necessary physical basis and verification for analytical and computational approaches to the development of engineering solutions. The members of SEM encompass a unique group of experimentalists, development engineers, design engineers, test engineers and technicians, and research and development scientists from industry and educational institutions working in materials; modeling and analysis; strain measurement and structural testing.

The Integrated Test Analysis Process for Structural Dynamic Systems

Robert N. Coppolino

ISBN: 978-3-031-79728-6    paperback
ISBN: 978-3-031-79729-3      ebook
ISBN: 978-3-031-79730-9    hardcover

DOI 10.1007/978-3-031-79729-3

A Publication in the Springer series
*SYNTHESIS SEM LECTURES ON EXPERIMENTAL MECHANICS*

Lecture #5
Series Editor: Kristin Zimmerman, *SEM*
Series ISSN
Print 2577-6053    Electronic 2577-6088

# The Integrated Test Analysis Process for Structural Dynamic Systems

Robert N. Coppolino
Measurement Analysis Corporation

*SYNTHESIS SEM LECTURES ON EXPERIMENTAL MECHANICS #5*

# ABSTRACT

Over the past 60 years, the U.S. aerospace community has developed, refined, and standardized an integrated approach to structural dynamic model verification and validation. One name for this overall approach is the Integrated Test Analysis Process (ITAP) for structural dynamic systems. ITAP consists of seven sequential tasks, namely: (1) definition of test article finite element models; (2) systematic modal test planning; (3) measured data acquisition; (4) measured data analysis; (5) experimental modal analysis; (6) systematic test-analysis correlation; and (7) reconciliation of finite element models and modal test data. Steps 1, 2, and 7 rely strictly on mathematical model disciplines, and steps 3 and 4 rely on laboratory disciplines and techniques. Current industry practice of steps 5 and 6 calls for interaction of mathematical model and laboratory disciplines, which compromises the objectivity of both modeling and laboratory disciplines. This book addresses technical content, strategies, and key relevant experiences related to all steps of ITAP, except for measured data acquisition which is the specialized domain of highly experienced laboratory professionals who contend with mechanical and electrical practicalities of instrumentation, excitation hardware, and data collection systems.

# KEYWORDS

model, test plan, data analysis, modal analysis, correlation, reconciliation

# Contents

# Preface

The Integrated Test Analysis Process (ITAP) for structural dynamic systems, presented in this book, offers in-depth expositions of six key process steps (recognizing the fact that measured data acquisition is the domain of highly specialized professionals, possibly requiring a completely separate book to properly cover that discipline).

Definition of an appropriate test article finite element model (FEM) involves (a) determination of the anticipated operational system's dynamic environments (specifically their frequency bandwidth and intensity envelopes). In the case of launch vehicles and spacecraft, NASA and USAF Space Command organizations specify typical bandwidths associated with dynamic environments (generally at or below 70 Hz for environments that do not include acoustic and shock load phenomena). Once the bandwidth of the dynamic environment is established, spatial resolution requirements for the test article FEM's structural components are defined based on upper frequency bound to wavelength relationships. As the FEM must be suited for adjustment of uncertain features, it must include definition of joints and component interfaces in a manner consistent with engineering drawings. The wealth of theoretical and experimental resources, especially for shell-type structures, offers opportunities to intelligently restrict attention to key sensitivity parameters.

Development of an effective modal test plan requires careful study of the test article's FEM predicted vibration modes. Specifically, close attention should be paid to analytical modal kinetic energy distributions in addition to modal frequencies and geometric mode shapes. Appreciation of the fact that modal kinetic energy is mathematically an "unpacking" of the mass-weighted orthogonality matrix points to the importance of modal kinetic energy to assist initial selection of response measurement (accelerometer) allocation. A common challenge associated with shell-type launch vehicle and spacecraft structures is the "many modes" problem that arises from the fact that many overall shell breathing modes occur in the same frequency band as overall body bending, torsion, and axial modes. Difficulties associated with this "many modes" problem may be alleviated by a well-informed target mode selection process based on modal decomposition of predicted flight dynamic events. Simplistic modal effective mass criteria, which employ target mode selection for restricted situations involving base excitation environments, are not relevant for more general situations. A definitive approach for allocation of response measurement (accelerometer) allocation is offered by the residual kinetic energy (RKE) method, which is widely employed in the U.S. aerospace community.

Preliminary measured data analysis employing a variety of metrics including probability density functions, autospectra, time history and associated spectrograms, and shock spectra is an invaluable prerequisite for detailed data analysis. Issues associated with anomalous data channels and unexpected (e.g., nonlinear) behavior can be noted and dealt with prior to engagement in detailed data analysis. Multiple Input/Multiple Output (MI/MO) spectral analysis procedures are the primary cornerstone for detailed measured data analysis. A cumulative coherence technique, based on Cholesky (triangular) decomposition of partially correlated excitation sources, provides a systematic tool for (1) assessment of the role (prominence) of individual excitations and (2) localization and characterization of nonlinear aspects of dynamic response (when prominent). The product of measured data analysis is estimated frequency response functions (FRFs) and accompanying coherence functions.

Experimental modal analysis is a discipline that benefits from techniques developed during the analog era (prior to 1971) and the digital era (after 1971). Intuitive graphical procedures for preliminary experimental modal analysis owe much of their content to analog era technology as well as newer procedures that highlight overall modal content (generally termed modal indicator functions); this represents the last opportunity for correction of problematic FRF data prior to detailed experimental modal analysis. A wide range of experimental modal analysis techniques have been developed during the post 1971 digital era. The techniques fall into two distinct categories, namely: (1) curve fitting procedures and (2) effective dynamic system estimation procedures. Simultaneous Frequency Domain (SFD) techniques belong to the latter category. Recent challenges encountered in NASA MSFC's Integrated Spacecraft and Payload Element (ISPE) modal test in 2016, associated with the "many modes" problem led to development of the SFD-2018 technique. This latest SFD innovation possesses a variety of features that alleviate the "many modes" challenge. Specifically, SFD-2018 validates estimated complex modes by a decoupling operation that defines single mode (SDOF equivalent) FRFs. This SFD-2018 operation is reminiscent of multi-shaker tuning, single mode isolation techniques developed during the analog era, without requiring multi-shaker tuning. In addition, since SFD-2018 automatically computes left-hand eigenvectors of an estimated state-space plant, the product of left- and right-hand eigenvector matrices automatically produces a mathematically perfect orthogonality matrix without reliance on a possibly flawed Test Analysis Mass (TAM) matrix. This feature of SFD-2018 alleviates common difficulties associated with satisfaction of both NASA STD-5002 and USAF Space Command SMC-S-004 test mode orthogonality criteria.

TAM mass matrix dependent test mode orthogonality and test-analysis cross-orthogonality criteria specified in NASA STD-5002 and USAF Space Command SMC-S-004 are widely used in the U.S. aerospace community. The most commonly employed strategy for systematic test analysis correlation involves employment of "real" experimental modes that are defined based on real mode curve fitting and/or approximate test modes constructed from the real component of estimated complex test modes. In most situations, this strategy is deemed appropriate. The NASA/MSFC ISPE modal test appears to present severe challenges to the commonly employed strategy. In response to this difficulty, a new complex test mode-based test-analysis cross-orthogonality procedure, which is independent of the TAM mass matrix, was developed. This provides further alleviation of difficulties presented by commonly employed cross-orthogonality criteria. Further analysis of ISPE test data suggests that ambiguities may occur as a result of employment of real test mode approximations. Specifically, complex mode orthogonality "unpacking" operations that produce test mode orthogonality distributions that are not necessarily in agreement with the commonly employed "real" test mode strategy. This difficulty led to introduction of an alternative "roadmap for a highly improved integrated test analysis process."

Real test mode-based reconciliation of FEMs and modal test data depends upon accurate, efficient parametric sensitivity analysis of the test article FEM and employment of robust modal cost functions. Augmentation of baseline model modes with a set of residual modes (called the residual mode augmentation (RMA) method) has been shown to eliminate unsatisfactory compromises inherent in a popular technique called structural dynamic modification (SDM). Employment of a balanced modal cost function was first successfully applied for test analysis reconciliation as part of the ISS P5 modal test conducted at NASA MSFC in 2001. It was determined at that time that Nelder–Meade Simplex optimization did not perform satisfactorily, while a Monte Carlo search strategy offered a robust search option. An exercise demonstrating hysteretic nonlinear system identification employing minimization of a time history based error norm is included as a final application. Nonlinearity and complex test modes are possibly the next challenges to be addressed in the continuing adventure called the integrated test analysis process for structural dynamic systems.

Robert N. Coppolino
October 2019

CHAPTER 1

# Overview of the Integrated Test Analysis Process (ITAP)

## 1.1    INTRODUCTION

Over the past 60 years, the U.S. aerospace community has developed, refined, and standard-ized an integrated approach to structural dynamic model verification and validation. One name for this overall approach is the Integrated Test Analysis Process (ITAP) for structural dynamic systems, which is summarized in Figure 1.1.

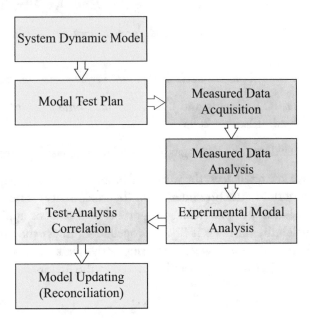

Figure 1.1: Overview of the integrated test analysis process.

The yellow-colored steps are associated with analytical-centric disciplines, while the aqua-colored steps are associated with laboratory-centric disciplines. The mixed-colored steps involve a blend of analytical and laboratory disciplines. Each step in the process is the product of a variety of authoritative, widely recognized sources. This book is intended to serve readers who have some familiarity with the foundations of structural dynamics and modal testing. A comprehensive treatment of the background subject matter for the seven steps is well beyond the scope of this book. In lieu of a thorough exposition of the seven ITAP steps, primary sources of their foundations are denoted herein.

## 1.2   SYSTEM DYNAMIC MODEL

System dynamics modeling rests on fundamental mathematical concepts in physics, particularly variational principles, structural mechanics, structural dynamics, matrix structural analysis, and the finite element method (FEM). Key references in these subject matters may be found in the following classic and modern publications.

1. R. B. Lindsay and H. Margenau, *Foundations of Physics*, Dover Publications, 1957.

2. Y. C. Fung, *Foundations of Solid Mechanics*, Prentice Hall, 1965.

3. S. P. Timoshenko, *History of Strength of Materials*, McGrawHill, 1953.

4. R. R. Craig and A. J. Kurdila, *Fundamentals of Structural Dynamics*, 2nd ed., Wiley, 2006.

5. H. Norman Abramson, *The Dynamic Behavior of Liquids in Moving Containers*, NASA-SP-106, 1966.

6. R. H. MacNeal, *Finite Elements: Their Design and Performance*, Marcel–Dekker, 1994.

7. O. C. Zienkiewicz, R. L. Taylor, and J. Z. Zhu, *The Finite Element Method, Its Basis and Fundamentals*, 6th ed., Elsevier, 2005.

8. L. Cremer, M. Heckl, and E. E. Ungar, *Structure Borne Sound*, Springer-Verlag, 1973.

   Among the above key references, the work of Abramson is particularly noteworthy in that a thorough exposition of fluid-structure interaction (hydroelasticity) is not treated in the other key references. Hydroelasticity is particularly important in aerospace applications, as roughly 80% of a launch vehicle's lift-off mass is liquid propellants (unless the propellants are solid). Also noteworthy is *Structure Borne Sound*, which singularly treats the empirical foundations of damping in structures (subject matter generally ignored in the literature).

## 1.3   MODAL TEST PLAN AND TEST-ANALYSIS CORRELATION

The modal test plan and test-analysis correlation, particularly in aerospace applications, are results of key contributions documented in many technical papers since 1960. Among the variety of background materials, the following references are noteworthy.

9. D. R. Martinez and A. Keith Miller, Ed., *Combined Experimental/Analytical Modeling of Dynamic Structural Systems*, ASME-AMD-67, 1985.

10. *Load Analysis of Spacecraft and Payloads*, NASA-STD-5002, 1996.

11. U.S. Air Force Space Command, *Independent Structural Loads Analysis*, SMC-S-004, 2008.

The 1985 book edited by Martinez and Miller documents the thought process at that time, which focused on hybrid analytical-experimental dynamic models. The "hybrid" approach focused on direct incorporation of experimental modal data (in lieu of reconciliation of experimental and analytical information), resulting in stringent requirements for modal test planning and test-analysis correlation that are incorporated in NASA-STD-5002 and SMC-S-004.

## 1.4    MEASURED DATA ACQUISITION

Measured data acquisition is a technical discipline and practice that is the domain of highly experienced specialists. While some reference materials are included in background literature in **Measured Data Analysis** and **General Sources**, the most comprehensive guidance is to be found in the community of experienced specialists.

## 1.5    MEASURED DATA ANALYSIS

The foundations of measured data analysis, in this book, are documented in the following books.

12. J. S. Bendat and A. G. Piersol, *Random Data Analysis and Measurement Procedures*, 4th ed., Wiley, 2010.

13. J. S. Bendat, *Nonlinear Systems Techniques and Applications*, 2nd ed., Wiley, 1998.

It should be noted that the sign convention employed by Bendat and Piersol for complex numbers and phase angle differ from most prevailing literature. The sign convention in the present book conforms to prevailing practice. Additional key references are listed in **General Sources**.

## 1.6    EXPERIMENTAL MODAL ANALYSIS

Experimental modal analysis has a rich historical background that is documented in three key sources, namely:

14. D. L. Brown and R. J. Allemang, The Modern Era of Experimental Modal Analysis, *Sound and Vibration Magazine*, January 2007.

15. R. E. D. Bishop and G. M. L. Gladwell, An Investigation into the Theory of Resonance Testing, *Philosophical Transactions, Royal Society of London, Series A*, 225(A-1055):241–280, 1963.

16. D. J. Ewins, *Modal Testing Theory and Practice*, Wiley, 1984.

That being said, the primary emphasis in this book's chapter on experimental modal analysis may be found in the following publications.

17. R. N. Coppolino, *A Simultaneous Frequency Domain Technique for Estimation of Modal Parameters from Measured Data*, SAE Paper 811046, 1981.

18. R. N. Coppolino, *Efficient and Enhanced Options for Experimental Mode Identification*, IMAC 21, 2003.

19. R. N. Coppolino, *Experimental Mode Verification (EMV) using Left-Hand Eigenvectors*, IMAC 37, 2019.

While the simultaneous frequency domain technique (SFD) is featured in this book, no implication as to its superiority over the variety of excellent experimental modal analysis techniques is implied.

## 1.7 MODEL UPDATING (RECONCILIATION)

The general subject of model updating (test-analysis reconciliation) is intimately tied to analytical approximation of structural dynamic model parametric sensitivity. It is most important to first cite techniques that are ultimately unreliable due to (a) the presence of closely spaced or repeated modes, (b) large uncertainties in structural joint (stiffness) parameters, and (c) substantial errors associated with modal truncation; the following published works describe those techniques in this category:

20. R. I. Fox and M. P. Kapoor, Rates of Change of Eigenvalues and Eigenvectors, *AIAA Journal*, 6, 1968.

21. R. B. Nelson, Simplified Calculation of Eigenvector Derivatives, *AIAA Journal*, 14, 1976.

22. P. Avitabile, *Twenty Years of Structural Dynamic Modification-A Review*, IMAC 20, 2002.

The technique that appears to efficiently and effectively circumvent deficiencies (a)–(c) is documented in the following.

23. R. Coppolino, *Methodologies for Verification and Validation of Space Launch System (SLS) Structural Dynamic Models*, NASA CR 2018-219800, 2018.

The reconciliation process is completed via incorporation of an optimization (error norm minimization) procedure. Three leading strategies designed to accomplish reconciliation are found in the following.

24. T. K. Hasselman, D. C. Zimmerman, and D. L. Herendeen, An Integrated FEA Software Capability for Dynamic Model Validation and Verification, *AIAA 40th SDM Conference*, 1999 (Bayesian Statistics).

25. J. A. Nelder and R. Mead, A Simplex Method for Function Minimization, *Computer Journal*, 7(4):308–313, 1965 (Gradient Search).

26. N. Metropolis, *The Beginning of the Monte Carlo Method*, Los Alamos Science (1987 Special Issue dedicated to Stanislaw Ulam), pp. 125–130 (Random Search).

## 1.8    GENERAL REFERENCES

Two additional key references that offer expositions related to most aspects of the Integrated Test Analysis Process for structural dynamic systems are as follows.

27. A. G. Piersol and T. L. Paez, Eds., *Harris' Shock and Vibration Handbook*, 6th ed., McGraw-Hill, 2010.

28. R. Allemang and P. Avitabile, Eds., *Handbook of Experimental Structural Dynamics*, Springer-Verlag, 2021.

## 1.9    KEY ILLUSTRATIVE EXAMPLES

Five key illustrative examples are employed throughout this book.

1. **Abramson Model:** An empty and fluid-filled circular cylindrical shell (modeled in closed form) serves to introduce (a) the "many modes" problem, (b) important parametric sensitivities, and (c) the profound effect of fluid mass on the frequency and content of structural dynamic modes.

2. **ISS P5 Modal Test:** End-to-end application of the integrated test analysis project completed in 2001 details (a) modal test planning, (b) measured data analysis, (c) experimental modal analysis, (d) test-analysis correlation, and test-analysis reconciliation, and (e) techniques for identification of nonlinear test article behavior.

3. **Axisymmetric Shell Structure:** A FEM is used to describe approaches for (a) understanding and characterizing modal dynamic characteristics, (b) selection of modal test target modes, and (c) evaluation of the effects of imperfections and local features (modal sensitivity) on otherwise axisymmetric systems.

4. **ISPE Modal Test:** This NASA/MSFC modal test program was completed in 2016. Independent review activities consisted of (a) development of a sensitivity analysis convergence test, (b) development of advanced experimental modal analysis methodology, and (c) formulation of a roadmap for a highly improved integrated test analysis process.

5. **Wire Rope Test:** Testing of a shock and vibration isolator test article, conducted as part of a U.S. Army Phase II SBIR ending in 2008, serves as an example for (a) measured data analysis and (b) system identification for a highly nonlinear dynamic system.

## 1.10   CLOSURE

This book is intended to provide details of the Integrated Test Analysis Process, principally from the perspective of the author's experience over the past 52 years. It presupposes the reader's basic knowledge of structural dynamics technology (engineering course work, hopefully supplemented by some graduate research and professional experience).

The previously cited references serve as essential recommended background materials. This book, to be sure, is a personal essay. It is certainly not intended to overshadow the excellent contributions of many individuals and organizations engaged in the field of analytical and experimental structural dynamics. The content of this book does not follow the usual logical sequence that is typical of foundational textbooks on structural dynamics and random data analysis. Instead, subject matters relevant to each of the seven key ITAP steps are presented in a manner that supports the content of each sequential chapter.

CHAPTER 2

# Definition of Test Article Finite Element Models

## 2.1 PART 1: VARIATIONAL FOUNDATIONS OF MODERN STRUCTURAL DYNAMICS

### 2.1.1 INTRODUCTION

The development of mathematical physics, of which structural dynamics is a branch, owes its present state to two general viewpoints. The first viewpoint due to Newton [1] sees nature following postulates describing the dynamic equilibrium of interacting bodies. The second viewpoint, due to d'Alembert [2], Hamilton [3], and Lagrange [4], sees nature following postulates describing efficient organization of energies (variational principles). The variational viewpoint, specifically Hamilton's principle, has guided the development of partial differential equations and natural boundary conditions for technical structural theories [5, 6].

Difficulties encountered in the quest for exact solutions of partial differential equations, subjected to natural boundary conditions, led to the introduction of approximate techniques based on Hamilton's principle. Ritz [7] employed assumed shape functions and generalized coordinates to deduce mass and stiffness matrices, which are foundational to the FEM [8] and Matrix Structural Analysis [9]. Galerkin [10] similarly applied assumed functions to a variational integral associated with a system's partial differential equations; his formulation, while influential in development of the FEM, has also been applied to solve nonlinear dynamic problems [11]. Trefftz [12] introduced a novel variational technique focusing on approximate satisfaction of natural boundary conditions, which are a byproduct of Hamilton's principle; his employment of shape functions that automatically satisfy the system's partial differential equations forms a basis of the boundary element method [13].

Hamilton's principle and the Ritz method, in particular, have been instrumental in the development of both the FEM and Matrix Structural Analysis. The basic building blocks of the FEM are most often developed on the basis of assumed boundary displacement referenced and optional interior displacement shape polynomials, which are used to define element mass and stiffness matrices based on the Ritz method. Assembly and analysis of structural system models falls in the category of Matrix Structural Analysis, which for a time before general acceptance of the FEM employed a force (degree of freedom) method.

Free vibration analysis of a matrix structural system is typically performed on an undamped system, for which the modes or real eigenvectors satisfy a mass weighted orthogonality relationship. Free vibration analysis of a damped structural system is rarely performed due to difficulties associated with theoretical damping matrices that do not relate to empirical modal damping. An alternative to the explicit damping matrix, namely a complex structural damping constant, is often employed in aeroelastic analysis [14]. The more widely used description of damping is based on critical damping ratio factors assigned to a truncated set of undamped system modes.

### 2.1.2  ECONOMY IN NATURE AND BASIC VARIATIONAL FORMULATIONS

Philosophers of antiquity, before the advent of modern science initiated by Newton (1686), believed that nature operated in accordance with a rule of economy as noted by Aristotle (d. 312 BC): "Nature follows the easiest path that requires the least amount of effort." In the medieval era, William of Ockham (1347) suggested an economic principle for human reasoning with his famous saying, "It is futile to employ many principles when it is possible to employ fewer" (popularly known as Ockham's razor). It is fascinating that Newton's second law may be used as a postulate to deduce variational formulations of mechanics (as theorems). Moreover, when the variational principle due to Hamilton (1824), is postulated, Newton's second law follows as a theorem.

The postulate-theorem interrelationship is summarized in Table 2.1 for a mechanical system composed of either (1) a single particle, i.e., single-degree-of-freedom (SDOF) system or (2) a multi-degree-of-freedom (MDOF) matrix system. Newton's second law states the foundational postulate, sequentially followed by d'Alembert's Principle and Hamilton's Principle, as "theorems."

Table 2.1: Hierarchal relationship of theoretical foundations of classical dynamics

| Theoretical Foundation | SDOF Systems | | MDOF (Matrix) Systems | |
|---|---|---|---|---|
| | Basic Equations | Supporting Information | Basic Equations | Supporting Information |
| Newton's 2nd Law (1689) | $F = \dfrac{d}{dt}\left(m\dfrac{du}{dt}\right)$ | | $[M]\{\ddot{U}\} = \{F\}$ | |
| d'Alembert's Principle (1743) | $\left(F - \dfrac{d}{dt}\left(m\dfrac{du}{dt}\right)\right) \cdot \delta u = 0$ | | $\{\delta U\}^T [M]\{\ddot{U}\} = \{\delta U\}^T \{F\}$ | |
| Hamilton's Principle (1824) | **Classical Statements** $\int_{t_1}^{t_2}\left(F - \dfrac{d}{dt}(m\dfrac{du}{dt})\right)\cdot\delta u\cdot dt = 0$ | **Integration by Parts** $\int_{t_1}^{t_2}\left(\delta\left(\dfrac{1}{2}m\left(\dfrac{du}{dt}\right)^2\right)+F\cdot\delta u\right)\cdot dt - \left.\left(m\dfrac{du}{dt}\right)\cdot\delta u\right\|_{t_1}^{t_2} = 0$ | **Classical Statements** $\int_{t_1}^{t_2}\{\delta U\}^T\{F - M\ddot{U}\}\cdot dt = \{0\}$ | **Integration by Parts** $\int_{t_1}^{t_2}\left(\delta\left(\dfrac{1}{2}\{\dot{U}\}[M]\{\dot{U}\}+\{\delta U\}^T\{F\}\right)\cdot dt - \{\delta U\}^T[M]\{\dot{U}\}\right\|_{t_1}^{t_2}=\{0\}$ |
| | $\int_{t_1}^{t_2}(\delta T + \delta W)\cdot dt = 0$ | **Kinetic Energy and Virtual Work** $T = \dfrac{1}{2}m\left(\dfrac{du}{dt}\right)^2 \quad \delta W = F\cdot\delta u$ | $\int_{t_1}^{t_2}(\delta T + \delta W)\cdot dt = 0$ | **Kinetic Energy and Virtual Work** $T = \dfrac{1}{2}\{\dot{U}\}^T[M]\{\dot{U}\} \quad \delta W = \{U\}^T\{F\}$ |
| | $\int_{t_1}^{t_2}(\delta L + \delta W)\cdot dt = 0$ | **Potential Energy, Non-Conservative Work, Lagrangian** $F = -\dfrac{dV}{du}+F_{NC}$ $\delta W \to F_{NC}\,\delta u$ $L = T - V$ | $\int_{t_1}^{t_2}(\delta L + \delta W)\cdot dt = 0$ | **Potential Energy, Non-Conservative Work, Lagrangian** $\{F\} = -\dfrac{d\{V\}}{d\{u\}}+\{F_{NC}\}F_{NC}$ $\delta W \to \{\delta U\}^T\{F_{NC}\}$ $L = T - V$ |

### 2.1.3    MATHEMATICAL PHYSICS AND HAMILTON'S PRINCIPLE

Application of Hamilton's principle to a dynamic system described as a continuum yields a volume integral of the type

$$\int_{t_1}^{t_2} \int_R (\delta T_R - \delta U_R + \delta W_R) \cdot dR \cdot dt = 0, \tag{2.1}$$

where $T_R$, $U_R$, and $\delta W_R$ are the kinetic energy, potential (or strain) energy, and virtual work functions per unit volume, $R$, respectively. Analysis of any particular dynamic system, described in terms of displacement variables, $u(x, y, z, t)$, which may be vectors, results in the following type of functionals:

$$\int_{t_1}^{t_2} \int_R (\text{P.D.E.}) \cdot \delta u \cdot dR \cdot dt + \int_{t_1}^{t_2} \int_S (\text{N.B.C.}) \cdot \delta u \cdot dS \cdot dt = 0, \tag{2.2}$$

"P.D.E." represents the particular partial differential equation(s) within the system's volume. "N.B.C." represents the natural boundary conditions, which are mathematically and physically admissible on the system's boundary surface(s), $S$. The general process for derivation of a system's partial differential equations and natural boundary conditions has provided a consistent basis for the development of technical structural theories for prismatic bars, beams, rings, plates, and shells.

### 2.1.4    THE CONTRIBUTIONS OF RITZ, GALERKIN, AND TREFFTZ

Three outstanding contributions that led to the development of approximate analysis techniques date back to the early part of the twentieth century. The methods bearing the names of Ritz, Galerkin, and Trefftz are all consequences of Hamilton's principle and the assumption of approximate solution functions.

### 2.1.5    THE RITZ METHOD

A monumental contribution to approximate analysis was introduced by Ritz (1908) [7], who described the displacement field in variable separable terms

$$u(x, y, z, t) = \sum_{n=1}^{N} \Psi_n(x, y, z) \cdot q_n(t), \tag{2.3}$$

where $\Psi_n(x, y, z)$ are assumed shape functions and $q_n(t)$ are temporal generalized coordinates (displacements). In addition, the strain field,

$$\varepsilon(x, y, z, t) = \sum_{n=1}^{N} \Psi_{\varepsilon,n}(x, y, z) \cdot q_n(t), \tag{2.4}$$

is linearly related to the assumed displacement field employing the appropriate partial derivatives. It should be noted that exact closed form solutions of partial differential equations are often expressed in variable separable form, whenever such solutions are possible. By assuming a series of functions that satisfy particular boundary conditions (or generally permit solution of natural boundary conditions), substitution of Equation (2.3) into Hamilton's principle (Equation 2.1), the following symmetric matrix equations are deduced:

$$[M]\{\ddot{q}\} + [K]\{q\} = [\Gamma]\{Q\}, \tag{2.5}$$

where the positive semi-definite, symmetric mass, and stiffness matrix terms are

$$M_{mn} = \int_R \text{``}\rho\text{''} \cdot \Psi_m \Psi_n dR, \qquad K_{mn} = \int_R \text{``}E\text{''} \cdot \Psi_{\varepsilon,m} \Psi_{\varepsilon,n} dR. \tag{2.6}$$

[Note: "$\rho$" and "$E$" are representative of mass density and elastic stiffness material properties.] In addition, the generalized forcing terms are governed by volume and surface integrals associated with applied body forces and surface loads, respectively.

The Ritz method, outlined above, was initially employed to approximately solve difficult problems described by partial differential equations and associated natural boundary conditions. Ultimately, it was extensively applied in development of the Finite Element Method [8] and Matrix Structural Analysis [9].

### 2.1.6   GALERKIN'S METHOD

Galerkin [10] defined an approximate method using the variable separable displacement field and associated generalized coordinates (Equations (2.3) and (2.4)) by substitution of the assumed functions into Equation (2.2) where the boundary conditions are automatically satisfied by choice of an appropriate set of spatial functions. The general statement of Galerkin's method is

$$\int_R (\text{P.D.E.}) \cdot \delta u \cdot dR = 0. \tag{2.7}$$

An appealing aspect of Galerkin's method is that it can be applied to any set of partial differential equations (even if a suitable variational formulation is unknown). The method has been successfully applied in the study of nonlinear dynamic systems [11].

### 2.1.7   TREFFTZ'S METHOD

Trefftz [12] proposed an approximate method that employs a set of assumed functions that automatically satisfy the partial differential equations. Therefore, the form of Equation (2.2) that must be satisfied is

$$\int_S (\text{N.B.C.}) \cdot \delta u \cdot dS = 0. \tag{2.8}$$

This method was generally ignored for about three decades until its advantage in the approximate solution of infinite domain problems was recognized. Trefftz's method has been instrumental in the development of the boundary element method.

## 2.1.8   AUTOMATED FORMULATIONS IN STRUCTURAL DYNAMICS

Among the methodologies that owe their foundations to Hamilton's principle and the contributions of Ritz, Galerkin, and Trefftz, two leading techniques are of prominence, namely (a) the Finite Element Method and (b) systematic Matrix Structural Analysis. Discussion of these two topics will be limited to linear systems.

## 2.1.9   THE FINITE ELEMENT METHOD

The Finite Element Method encompasses three mathematical disciplines, namely: (1) definition of building blocks (finite elements), (2) assembly of structural system models, and (3) solution of structural system equations. The third discipline is addressed by systematic Matrix Structural Analysis.

Finite elements are generally derived as a specialized application of the Ritz method. The displacement field associated with a single element is most generally described in terms of the following family of shape functions:

$$u(x, y, z, t) = \sum_{n=1}^{N_h} \Psi_{h,n}(x, y, z) \cdot u_n(t) + \sum_{n=1}^{N_p} \Psi_{p,n}(x, y, z) \cdot q_n(t)$$

(2.9)

or in matrix form,

$$\{u\} = [\Psi_h] \{u_h\} + [\Psi_p] \{q_p\}.$$

The "$h$" shape functions are referenced to physical displacements at specific grid points on the element's (boundary) surface and the "$p$" shapes are polynomial functions that have null value along the element's boundary surface. Substitution of the above shape family functions (and their appropriate strain partial derivatives) into Hamilton's principle (Equation 2.1) results in positive semi-definite mass and stiffness matrices of the forms

$$[M]_{\text{element}} = \begin{bmatrix} M_{pp} & M_{ph} \\ M_{hp} & M_{hh} \end{bmatrix}_{\text{element}}, \qquad [K]_{\text{element}} = \begin{bmatrix} K_{pp} & K_{ph} \\ K_{hp} & K_{hh} \end{bmatrix}_{\text{element}}.$$

(2.10)

It should be emphasized here that definition of well-posed and accurate finite elements depends upon expert selection of shape functions and numerical integration schemes. The most commonly employed finite elements in commercial finite element codes is the "$h$" element which does not include "$p$" generalized coordinates. The more general elements are often called "$h$-$p$" elements.

## 2.1.10   ASSEMBLY OF LINEAR STRUCTURAL SYSTEM MODELS

Assembly of a linear system structural dynamic model involves the allocation and superposition (overlapping) of individual finite elements onto a system degree-of-freedom (DOF) map. This process defines sparse system mass and stiffness matrices that are positive semi-definite and symmetric. A system composed of an assembly of "$h$" elements is defined by "grid" set mass and stiffness matrices denoted by $[M]$ and $[K]$, respectively. Additional quantities complete the ingredients for a structural dynamic model, namely: (1) allocation of excitation forces to system grid points, (2) formation of an assumed viscous damping matrix (which unfortunately does not resemble physical reality in most commercial finite element codes), and (3) allocation of local nonlinear internal forces. The grid set equations for the structural dynamic model are of the for

$$[M]\{\ddot{U}\} + [B]\{\dot{U}\} + [K]\{U\} = [\Gamma_e]\{F_e\} + [\Gamma_N]\{F_N(U_N, \dot{U}_N, p)\},$$
$$\{U_N\} = [\Gamma_N]^T\{U\}, \{\dot{U}_N\} = [\Gamma_N]^T\{\dot{U}\}. \tag{2.11}$$

Constraints and boundary conditions, collectively described by transformations of the form,

$$\{u\} = [G]\{u_a\}, \tag{2.12}$$

are applied in a symmetric manner as a consequence of the quadratic forms and integrals defined by the Ritz method, resulting in the "analysis" set equations and matrices (for a linear system)

$$[M_{aa}]\{\ddot{u}_a\} + [B_{aa}]\{\dot{u}_a\} + [K_{aa}]\{u_a\} = [\Gamma_{ae}]\{F_e\}, \tag{2.13}$$

where the reduced order matrices are

$$[M_{aa}] = \left[G_{ga}^T M_{gg} G_{ga}\right], \qquad [B_{aa}] = \left[G_{ga}^T B_{gg} G_{ga}\right],$$
$$[K_{aa}] = \left[G_{ga}^T K_{gg} G_{ga}\right], \qquad [\Gamma_a] = \left[G_{ga}\right]^T [\Gamma_g]. \tag{2.14}$$

## 2.1.11   SYSTEM MODELS WITH LOCALIZED NONLINEARITIES

Structural assemblies sometimes display locally nonlinear behavior (primarily at structural joints that connect subassemblies). The "analysis" set equations for such situations are augmented by localized nonlinear forces, $\{F_N\}$, allocated in accordance with the nonlinear force distribution matrix, $[\Gamma_{aN}]$, are of the form

$$[M_{aa}]\{\ddot{u}_a\} + [B_{aa}]\{\dot{u}_a\} + [K_{aa}]\{u_a\} = [\Gamma_{ae}]\{F_e\} + [\Gamma_{aN}]\{F_N\}. \tag{2.15}$$

The nonlinear forces are defined as functions of localized displacements and velocities that are defined as

$$\{u_N\} = \left[\Gamma_{aN}^T\right]\{u_a\}, \qquad \{\dot{u}_N\} = \left[\Gamma_{aN}^T\right]\{\dot{u}_a\}. \tag{2.16}$$

In the most general case, localized nonlinear forces are "hysteretic" functions of localized displacements, velocities, and (path dependent) parameters, $\{p\}$, according to algorithms of the following general form:

$$\{F_N(t), p(t + dt)\} = \{f(u_N(t), \dot{u}_N(t), p(t)\}. \tag{2.17}$$

For the more restricted cases in which the parameters are not path dependent, the localized nonlinear forces are "algebraic," rather than "hysteretic."

## 2.1.12   MATRIX STRUCTURAL ANALYSIS

Matrix Structural Analysis predates the Finite Element Method by several decades. Historically, matrix formulations for structural systems developed along two paths, namely (a) the force method and (b) the displacement method. The displacement method, based primarily on Hamilton's principle and the Ritz method, ultimately eclipsed the force method due to the advent of the finite element method. Matrix structural analysis procedures, applied to damped systems that include localized nonlinear forces, most often employ an efficient transformation defined by free vibration modal analysis of an undamped system. [Note: In the following exposition on modal analysis, the "$a$" set subscript employed in Equations (2.12)–(2.16) is dropped.]

## 2.1.13   FREE VIBRATION AND MODAL ANALYSIS

Free vibration of an undamped structural system is described by solutions of the type

$$\{u(t)\} = \{\Phi_n\} \cdot \sin(\omega_n t) \quad \text{or using complex exponentials} \quad \{u(t)\} = \{\Phi_n\} \cdot e^{i\omega_n t}, \tag{2.18}$$

which define the real eigenvalue problem

$$[K]\{\Phi_n\} - [M]\{\Phi_n\}\lambda_n = \{0\} \qquad (\lambda_n = \omega_n^2), \tag{2.19}$$

which has as many independent eigenvectors, $\{\Phi_n\}$ (normal modes) and eigenvalues ($\lambda_n = \omega_n^2$), as the system matrix order. The collection of all or a truncated subset of normal modes (the modal matrix) defines the modal transformation

$$\{u\} = [\Phi]\{q\}. \tag{2.20}$$

The modal matrix has the following mathematical properties (for unit mass normalized modes):

$$[\Phi]^T [M][\Phi] = [I], \qquad [\Phi]^T [K][\Phi] = [\lambda], \tag{2.21}$$

where $[\lambda]$ is a diagonal matrix of eigenvalues.

## 2.1.14   UNCOUPLED STRUCTURAL DYNAMIC EQUATIONS

Application of the modal transformation to Equations (2.15) and (2.16) results in

$$\{\ddot{q}\} + [\Phi^T B \Phi]\{\dot{q}\} + [\lambda]\{q\} = [\Phi^T \Gamma_e]\{F_e\} + [\Phi^T \Gamma_N]\{F_N\},$$
$$\{u_N\} = [\Gamma_N^T \Phi]\{q\}, \qquad \{\dot{u}_N\} = [\Gamma_N^T \Phi]\{\dot{q}\}. \tag{2.22}$$

In general, the "modal" damping matrix, $\left[\Phi^T B \Phi\right]$, is fully coupled. However, an uncoupled approximation (based on a wide variety of experimental results) is quite common, resulting in the fully uncoupled modal dynamic equations,

$$\ddot{q}_n + 2\zeta_n \omega_n \dot{q}_n + \omega_n^2 q_n = \left[\Phi_n^T \Gamma_N\right]\{F_N\} + \left[\Phi_n^T \Gamma_e\right]\{F_e\}. \tag{2.23}$$

### 2.1.15   CLOSURE

The theoretical foundations of finite element analysis and matrix structural analysis, which rely on systematic variational principles, have been outlined. Free vibration modes of undamped, linear structural dynamic systems have been shown to uncouple general MDOF structural dynamic matrix equations with local nonlinear features.

## 2.2   PART 2: GUIDELINES FOR SYSTEMATIC MODEL DEVELOPMENT

### 2.2.1   INTRODUCTION

Definition of the relevant structural dynamic model of a vehicle or stationary system begins with determination of the frequency band associated with anticipated dynamic service environments. Shock and response spectra [15] offer a general approach for estimation the frequency band.

### 2.2.2   LINEAR SINGLE-DEGREE-OF-FREEDOM EQUATIONS

An elementary linear structural dynamic or mechanical dynamic system is described in terms of a SDOF linear mechanical system. The dynamic response, $u(t)$, of a SDOF system which is excited by an applied force, $F(t)$, and/or base (foundation) motion input, $u_0(t)$, is governed by the ordinary differential equation (schematic depicted in Figure 2.1),

$$m\ddot{u}(t) + b\dot{u}(t) + ku(t) = F(t) + b\dot{u}_0(t) + ku_0(t), \tag{2.24}$$

where $m, b$, and $k$ are the constant mass, viscous damping, and elastic stiffness coefficients, respectively. By defining the relative displacement variable, $u_R(t) = u(t) - u_0(t)$, and dividing by the mass, $m$, Equation (2.1) simplifies to

$$\ddot{u}_R(t) + 2\zeta_n \omega_n \dot{u}_R(t) + \omega_n^2 u_R(t) = F(t)/m - \ddot{u}_0(t), \tag{2.25}$$

where the undamped natural frequency (rad/sec) and critical damping ratio, respectively, are

$$\omega_n = \sqrt{\frac{k}{m}}, \qquad \zeta_n = b/(2m\omega_n). \tag{2.26}$$

In addition, the damped natural frequency (rad/sec) is defined as

$$\omega_d = \omega_n \sqrt{1 - \zeta_n^2}. \tag{2.27}$$

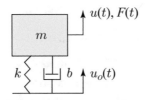

Figure 2.1: Linear SDOF dynamic system schematic.

### 2.2.3    LINEAR MULTI-DEGREE-OF-FREEDOM EQUATIONS

Linear structural dynamic systems are generally described in terms of partial differential equations of mathematical physics or matrix differential equations constructed as assemblies of finite elements. The typical form for matrix (MDOF) structural dynamic equations is

$$[M]\{\ddot{U}(t)\} + [B]\{\dot{U}(t)\} + [K]\{U(t)\} = [\Gamma_e]\{F_e(t)\}. \tag{2.28}$$

Symmetric system mass, damping, and stiffness matrices, $[M]$, $[B]$, and $[K]$, respectively, are associated with discrete displacement degrees of freedom, $\{U(t)\}$. The external loads, $\{F_e(t)\}$ are allocated to the discrete displacements in accordance with geometric distributions described by the (linearly independent) columns of the matrix, $[\Gamma_e]$.

By partitioning the displacements into "interior" and "boundary" subsets,

$$\{U\} = \left\{ \begin{array}{c} U_i \\ U_b \end{array} \right\}, \tag{2.29}$$

the MDOF equations are expressed as

$$\left[ \begin{array}{cc} M_{ii} & M_{ib} \\ M_{bi} & M_{bb} \end{array} \right] \left\{ \begin{array}{c} \ddot{U}_i \\ \ddot{U}_b \end{array} \right\} + \left[ \begin{array}{cc} B_{ii} & B_{ib} \\ B_{bi} & B_{bb} \end{array} \right] \left\{ \begin{array}{c} \dot{U}_i \\ \dot{U}_b \end{array} \right\}$$
$$+ \left[ \begin{array}{cc} K_{ii} & K_{ib} \\ K_{bi} & K_{bb} \end{array} \right] \left\{ \begin{array}{c} U_i \\ U_b \end{array} \right\} = \left[ \begin{array}{c} \Gamma_{ie} \\ \Gamma_{be} \end{array} \right] \{F_e\}. \tag{2.30}$$

The transformation of "interior" displacements to "relative" displacements with respect to the "boundary" displacements is defined as

$$\left\{ \begin{array}{c} U_i \\ U_b \end{array} \right\} = \left[ \begin{array}{cc} I_{ii} & -K_{ii}^{-1}K_{ib} \\ 0_{bi} & I_{bb} \end{array} \right] \left\{ \begin{array}{c} U_i' \\ U_b \end{array} \right\}. \tag{2.31}$$

Symmetric application of this transformation to Equation (2.30) results in

$$\left[ \begin{array}{cc} M_{ii} & M'_{ib} \\ M'_{bi} & M'_{bb} \end{array} \right] \left\{ \begin{array}{c} \ddot{U}'_i \\ \ddot{U}_b \end{array} \right\} + \left[ \begin{array}{cc} B_{ii} & B'_{ib} \\ B'_{bi} & B'_{bb} \end{array} \right] \left\{ \begin{array}{c} \dot{U}'_i \\ \dot{U}_b \end{array} \right\}$$
$$+ \left[ \begin{array}{cc} K_{ii} & 0'_{ib} \\ 0'_{bi} & K'_{bb} \end{array} \right] \left\{ \begin{array}{c} U'_i \\ U_b \end{array} \right\} = \left[ \begin{array}{c} \Gamma_{ie} \\ \Gamma'_{be} \end{array} \right] \{F_e\}. \tag{2.32}$$

The interior partition of the transformed structural dynamics equations is

$$[M_{ii}]\{\ddot{U}'_i\} + [B_{ii}]\{\dot{U}'_i\} + [K_{ii}]\{U'_i\} = [\Gamma_{ie}]\{F_e\} - [M_{ib}]\{\ddot{U}_b\} - [B_{ib}]\{\dot{U}_b\}. \quad (2.33)$$

Note the similarity of form of SDOF (Equation (2.25)) and MDOF (Equation (2.33)) "relative" displacement relationships (with the exception of the MDOF damping term, $[B_{ib}]$). It should be noted that for a rigid foundation, $[B_{ib}]$ vanishes.

## 2.2.4 NORMAL MODES OF UNDAMPED MDOF SYSTEMS

Free vibration of an undamped, linear structural dynamic system is described by solutions of the real eigenvalue problem (retaining the notation of Equation (2.33)),

$$[K_{ii}]\{\Phi_{in}\} - [M_{ii}]\{\Phi_{in}\}\lambda_n = \{0\}, \qquad (\lambda_n = \omega_n^2), \quad (2.34)$$

where $\{\Phi_{in}\}$ are distinct, orthogonal individual eigenvectors (or mode shapes), and $\lambda_n = \omega_n^2$, are the corresponding eigenvalues (note: $\omega_n$ are circular natural frequencies). The collection of all (or a truncated set of) eigenvectors, $[\Phi_i]$, defines the real mode displacement transformation,

$$\{U_i(t)\} = [\Phi_i]\{q(t)\}, \quad (2.35)$$

which has the following decoupling mathematical properties (for unit mass normalized modes):

$$[\Phi_i]^T [M_{ii}][\Phi_i] = [I], \qquad [\Phi_i]^T [K_{ii}][\Phi_i] = [\lambda]. \quad (2.36)$$

Note that the terms of the diagonal matrix, $[\lambda]$, are the real eigenvalues, $\lambda_n = \omega_n^2$.

Application of the real mode displacement transformation to Equation (2.34), results in the modal equations

$$\{\ddot{q}\} + [\Phi_i^T B_{ii} \Phi_i]\{\dot{q}\} + [\lambda]\{q\}$$
$$= [\Phi_i^T \Gamma_{ie}]\{F_e\} - [\Phi_i^T M_{ib}]\{\ddot{U}_b\} - [\Phi_i^T B_{ib}]\{\dot{U}_b\}. \quad (2.37)$$

A common approximation for the "modal" damping matrix assumes it is "uncoupled" (or diagonal). Moreover, the right side damping term, $[\Phi_i^T B_{ib}]$ is often assumed negligible, resulting in the uncoupled modal equations, which are compared directly to SDOF Equation (2.25) as follows:

$$\underline{\ddot{q}_n(t)} + \underline{2\zeta_n\omega_n\dot{q}_n(t)} + \underline{\omega_n^2 q_n(t)} = \underline{[\Phi_{in}^T \Gamma_{ie}]\{F_e(t)\}} - [\Phi_{in}^T M_{ib}]\{\ddot{U}_b(t)\},$$
$$\downarrow \qquad \downarrow \qquad \downarrow \qquad \downarrow \qquad \nearrow \qquad (2.38)$$
$$\ddot{u}_R(t) + 2\zeta_n\omega_n\dot{u}_R(t) + \omega_n^2 u_R(t) = F(t)/m - \ddot{u}_0(t).$$

While many more consequences of the modal transformation will be discussed throughout this book, the clear relationship between linear SDOF system dynamics and uncoupled modal dynamics (Equation (2.37)) will be exploited in the following sections for systematic definition of MDOF dynamic system modeling requirements.

### 2.2.5   RESPONSE TO IMPULSIVE AND TRANSIENT FORCE EXCITATIONS

Consider the case of a linear SDOF system excited by applied loads,

$$m\ddot{u}(t) + b\dot{u}(t) + ku(t) = F(t) \quad \text{or} \quad \ddot{u}(t) + 2\zeta_n\omega_n\dot{u}(t) + \omega_n^2 u(t) = F(t)/m. \qquad (2.39)$$

A pure impulsive force, $\hat{F}$, is defined as the finite integral,

$$\hat{F} = \int_{0-}^{0+} F(t)dt, \qquad (2.40)$$

over which the force approaches infinity as the time interval approaches zero duration. When such a loading is applied to a system, which is initially at rest, the velocity response immediately after the impulse ($t = 0+$) is

$$\dot{u}(0+) = \hat{F}/m, \qquad (2.41)$$

and the ensuing free decay response is

$$m\omega_n u(t)/\hat{F} = h(t) = e^{-\zeta_n\omega_n t}\sin(\omega_d t), \quad \text{where} \quad \omega_d = \sqrt{1 - \zeta_n^2}\,\omega_n. \qquad (2.42)$$

Response of the SDOF system to an impulsive force, $\hat{F}(\tau) = F(\tau)d\tau$, is the "particular" solution,

$$\begin{aligned}
u_p(t) &= 0 & \text{for } t < \tau \\
u_p(t) &= (1/m\omega_n)\,h(t - \tau)F(\tau)d\tau & \text{for } t \geq \tau.
\end{aligned} \qquad (2.43)$$

This function forms the basis of the Duhamel integral formula for SDOF system response to a general transient force excitation [15], which is the summation or integral of responses to a continuous train of pulses,

$$u_p(t) = \int_0^t (1/m\omega_n)\,h(t - \tau)F(\tau)d\tau. \qquad (2.44)$$

For the more general case of a linear SDOF system with initial "steady-state" displacement associated with an applied force that begins with a finite value, $F(0)$, an additional "homogeneous" solution component is required to eliminate a potentially dominant transient. Specifically, the complete response is

$$u(t) = \int_0^t (1/m\omega_n)\,h(t - \tau)F(\tau)d\tau + F(0)/\left(m\omega_n^2\right) \cdot e^{-\zeta_n\omega_n t}\cos(\omega_d t). \qquad (2.45)$$

## 2.2.6   RESPONSE SPECTRUM AND SHOCK SPECTRUM

Two closely related "signature" functions for linear SDOF system response to transient excitation are the response spectrum and shock spectrum [15]. The response spectrum is associated with response of a unit mass SDOF system excited by an applied force, $F(t)$, in particular,

$$\ddot{u}(t) + 2\zeta_n\omega_n\dot{u}(t) + \omega_n^2 u(t) = F(t). \tag{2.46}$$

The maximum and minimum response map linear SDOF systems to a specific force history, $F(t)$, with selected value of $\zeta_n$ for the range of natural frequency, $0 < \omega_n < \omega_{max}$ is defined as the **response spectrum**. The response spectrum is defined for extremes in displacement, velocity and acceleration response. Normalized response spectra are defined (using $F_0 = \max(|F(t)|)$) as follows:

$$RD\left(\omega_n, \zeta_n, F(t)\right) = \left[\max\left(\omega_n^2 u(t)/F_0\right),\ \min\left(\omega_n^2 u(t)/F_0\right)\right]$$
$$RV\left(\omega_n, \zeta_n, F(t)\right) = \left[\max\left(\omega_n\dot{u}(t)/F_0\right),\ \min\left(\omega_n\dot{u}(t)/F_0\right)\right] \tag{2.47}$$
$$RA\left(\omega_n, \zeta_n, F(t)\right) = \left[\max\left(\ddot{u}(t)/F_0\right),\ \min\left(\ddot{u}(t)/F_0\right)\right].$$

The shock spectrum is associated with response of a linear SDOF system excited by an applied base acceleration, in particular,

$$\ddot{u}_R(t) + 2\zeta_n\omega_n\dot{u}_R(t) + \omega_n^2 u_R(t) = -\ddot{u}_0(t), \tag{2.48}$$

$$u_R(t) = u(t) - u_0(t). \tag{2.49}$$

The maximum and minimum response map SDOF systems to a specific base acceleration history, $\ddot{u}_0(t)$, with selected value of $\zeta_n$ for the range of natural frequency, $0 < \omega_n < \omega_{max}$ is defined as the **shock spectrum**. The shock spectrum is defined for extremes in relative displacement, relative velocity and absolute acceleration response. Normalized response spectra are defined (using $\ddot{U}_0 = \max\left(|\ddot{u}_0(t)|\right)$) as follows:

$$SD\left(\omega_n, \zeta_n, \ddot{u}(t)\right) = \left[\max\left(\omega_n^2 u_R(t)/\ddot{U}_0\right),\ \min\left(\omega_n^2 u_R(t)/\ddot{U}_0\right)\right]$$
$$SV\left(\omega_n, \zeta_n, \ddot{u}(t)\right) = \left[\max\left(\omega_n\dot{u}_R(t)/\ddot{U}_0\right),\ \min\left(\omega_n\dot{u}_R(t)/\ddot{U}_0\right)\right] \tag{2.50}$$
$$SA\left(\omega_n, \zeta_n, \ddot{u}(t)\right) = \left[\max\left(\ddot{u}(t)/\ddot{U}_0\right),\ \min\left(\ddot{u}(t)/\ddot{U}_0\right)\right].$$

## 2.2.7   ILLUSTRATIVE EXAMPLE: HALF-SINE PULSE FORCE

Consider the half-sine pulse force excitation, shown in Figure 2.2. The normalized displacement **response spectrum** (assuming $\zeta_n = 0.02$) for this excitation is also shown in Figure 2.2.

A more commonly employed plot of response spectrum is based on the peak absolute values of the function, as depicted in Figure 2.3.

A useful property of the absolute displacement response spectrum is that its asymptotic value, for frequencies above $f^*$ approaches unity. The physical significance of the cut-off frequency, $f^*$, is that linear SDOF systems with natural frequency, $f_n > f^*$, respond "quasi-statically" to the excitation environment, $F(t)$, i.e., the excitation is "slowly varying" for a system

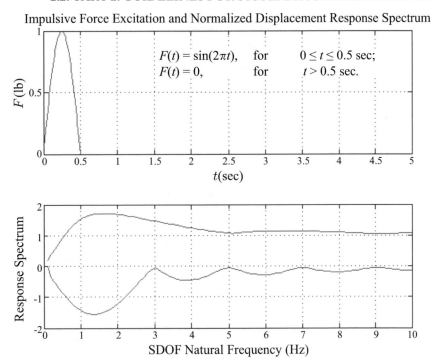

Figure 2.2: Half-sine pulse and normalized displacement response spectrum.

with relatively high natural frequency. This characteristic also applies to relative displacement and absolute acceleration shock spectra.

### 2.2.8    ILLUSTRATIVE EXAMPLE: 1940 EL CENTRO CA GROUND MOTION DATA

A measured ground motion time history recorded during May 18, 1940 El Centro CA earthquake is illustrated in Figure 2.4.

The normalized response spectrum (computed for $\zeta_n = 0.02$), provided in Figure 2.5, suggests a cut-off frequency, $f^* = 20$ Hz for this particular event.

### 2.2.9    DEFINITION OF RELEVANT STRUCTURAL DYNAMIC MODELS

In order to develop a relevant dynamic model, general requirements should be addressed based on:

(1)  frequency band, $0 < f < f^*$, and intensity ($F_0$) of anticipated dynamic environments; and

(2)  general characteristics of structural or mechanical components.

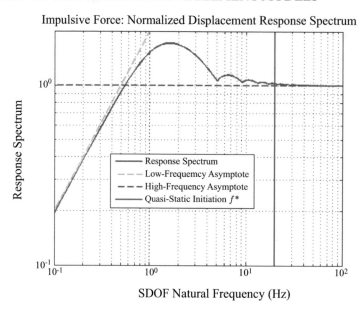

Figure 2.3: Absolute normalized displacement response spectrum.

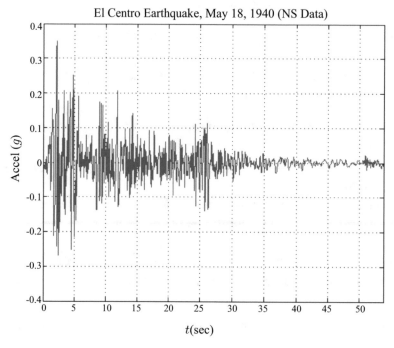

Figure 2.4: El centro NS acceleration time history.

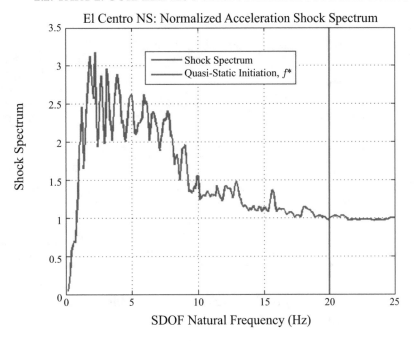

Figure 2.5: El centro NS acceleration normalized response spectrum.

With the cut-off frequency ($f^*$) established based on shock and/or response spectra of anticipated dynamic environments, the shortest relevant wavelength of forced vibration for components in a structural assembly may be calculated. For finite element modeling, the quarter wavelength ($L/4$) is of particular interest, since it defines the grid spacing requirement needed to accurately model dynamics [15]. Table 2.2 summarizes guidelines for typical structural components.

In addition to the above grid spacing guidelines, the engineer must also consider limitations associated with beam and plate technical theories. In particular, if the wavelength to thickness ratio ($L/h$) is less than about 10, a higher-order theory (than pure flexure) or 3-D elasticity modeling should be considered [16]. Moreover, modeling requirements for capture of stress concentration details may call for finer grid meshing than suggested by the cut-off frequency. Finally, if the dynamic environment is of sufficiently high amplitude ($F_0$), nonlinear modeling may be required, e.g., if plate deflections are greater than thickness, $h$ [17].

## 2.2.10   ILLUSTRATIVE EXAMPLE: ALUMINUM LAUNCH VEHICLE AND SPACECRAFT COMPONENTS

Consider a cut-off frequency, $f^* = 50$ Hz, which is typical for launch vehicle and spacecraft primary structural loads. Assuming the structural components are aluminum ($E \sim 10^7$ psi,

Table 2.2: Quarter wavelength relationships for typical structural components

| Basis | Component | Wave Type | L/4 | Additional Data |
|---|---|---|---|---|
| Continuum Mechanics | 3-D Elastic | Dilational | $(\sqrt{E/\rho})/(4f^*)$ | $E$ = stretch modulus $G$ = shear modulus $B$ = bulk modulus $\rho$ = mass density |
| | | Shear | $(\sqrt{G/\rho})/(4f^*)$ | |
| | 3-D Acoustic | Dilational | $(\sqrt{B/\rho})/(4f^*)$ | |
| Technical Theory | String | Lateral | $(\sqrt{T/\rho A})/(4f^*)$ | $T$ = tension $A$ = cross-sectional area $EI$ = flexural stiffness |
| | Rod | Axial | $(\sqrt{E/\rho})/(4f^*)$ | |
| | | Torsion | $(\sqrt{G/\rho})/(4f^*)$ | |
| | Beam | Bending | $(\pi/2)(EI/\rho A)^{1/4}/\sqrt{2\pi f^*}$ | |
| | Membrane | Axial | $(\sqrt{N/\rho h})/(4f^*)$ | $h$ = plate thickness $D$ = plate flexural stiffness $N$ = in-plane stress resultant |
| | Plate | Bending | $(\pi/2)(D/\rho h)^{1/4}/\sqrt{2\pi f^*}$ | |

$G \sim 3.84 \times 10^6$ psi, $\rho \sim 2.59 \times 10^{-4}$ lb-sec²/in⁴), the following quarter wavelengths ($L/4$) are defined for continuum mechanics theory:

a. Dilational deformation: $L/4 = \sqrt{E/\rho}/(4f^*) \sim 982$ in.

b. Shear deformation: $L/4 = \sqrt{G/\rho}/(4f^*) \sim 609$ in.

Clearly, the cross-sectional dimension of rod, beam, and plate-shell components of the largest feasible launch vehicle and spacecraft structures are substantially smaller than the above $L/4$ estimates. Therefore, models based on technical theories are most feasible.

## 2.2.11   ILLUSTRATIVE EXAMPLE: ALUMINUM LAUNCH VEHICLE FEEDLINE

Consider a large-scale launch vehicle's propellant feedline with the following cross-sectional dimensions (not representative of any particular system):

Outer diameter:   $OD = 15$  in, wall thickness:   $h = 0.5$  in.

The feedline's cross-sectional parameters are $A = 22.7$ in², $I = 599$ in⁴. Therefore, employing the relevant technical theory formulae, the $L/4$ estimates are:

c. Axial deformation: $L/4 = \sqrt{E/\rho}/(4f^*) \sim 982$ in.

d. Torsional deformation: $L/4 = \sqrt{G/\rho}/(4f^*) \sim 609$ in.

In order to properly consider lateral bending dynamics of the propellant feedline, the contained fluid must be considered. For the purposes of the present discussion, water is employed as an ersatz fluid with properties in the general range of liquid oxygen (LOX); the effective bulk modulus of water at room temperature, accounting for radial flexibility of the feedline structure is $B \sim 1.6 \times 10^5$ psi, and its mass density is $\rho \sim 1.36 \times 10^{-5}$ lb-sec$^2$/in$^4$. On the basis of continuum theory, the quarter wavelength $(L/4)$ for the fluid within the feedline is:

e. Dilational fluid deformation: $L/4 = \sqrt{B/\rho}/(4f^*) \sim 207$ in.

Clearly, the cross-sectional dimension of the feedline is an order of magnitude less than $L/4$. Therefore, a fluid model corresponding to a structural "rod" enclosed within the feedline structure, yet permitted to slide with respect to the rod, is most appropriate.

Since the contained fluid is constrained to move laterally with the feedline structure, the "lateral" mass per unit length, $\rho A \sim .0213$ lb-sec$^2$/in$^2$, is the sum of structural and fluid components. Therefore, employing the relevant technical (Euler–Bernoulli) beam theory, the lateral $L/4$ estimate for the feedline is

f. Feedline bending deformation: $L/4 = (\pi/2) \cdot (EI/\rho A)^{1/4} / \sqrt{4f^*} \sim 80.8$ in.

This result indicates that the ratio of quarter wavelength to feedline diameter is roughly 5 at $f^* = 50$ Hz, suggesting that a feedline structural model accounting for cross-sectional shear deformation (Timoshenko beam theory) may be most appropriate. Fortunately, most modern FEM codes employ beam elements that are based on Timoshenko beam theory.

### 2.2.12 MODAL DENSITY AND THE EFFECTIVENESS OF FINITE ELEMENT MODELS

Finite element modeling is an effective approach for study of structural and mechanical system dynamics as long as individual vibration modes have "sufficient frequency spacing" or "low modal density." Modal density is typically described as the number of modes within a 1/3 octave frequency band ($f_0 < f < 1.26 f_0$). When modal density of a structural component or structural assembly is greater than 10 modes per 1/3 octave band, details of individual vibration modes are not of significance and statistical vibration response characteristics are of primary importance. In such a situation, the Statistical Energy Analysis (SEA) method [18] applies.

Table 2.3 [15] gives formulae for modal density (as a mathematical derivative, $dn/d\omega$ ($n$ = number of modes, $\omega$ = frequency in radians/sec), for typical structural components.

The above modal density relationships relate to structural components, rather than structural assemblies that are composed of a variety of components. More reliable estimation of a particular system's modal density is based on the assembled system's modes.

### 2.2.13 ILLUSTRATIVE EXAMPLE: FLUID-FILLED CIRCULAR CYLINDRICAL SHELL

The following example system, consisting of a fluid-filled, thin-walled circular cylindrical shell (see Figure 2.6), was extensively studied by Abramson [19] during the mid-1960s. It continues to offer excellent, comprehensive guidance for modeling of launch vehicle propellant tank

Table 2.3: Modal densities of typical structural components

| Component | Motion | Modal Density, $dn/d\omega$ | Additional Data |
|-----------|--------|----------------------------|-----------------|
| String | Lateral | $L/(\pi\sqrt{T/\rho A})$ | $T$ = tension, $A$ = area $\rho$ = mass density $L$ = length |
| Rod | Axial | $L/(\pi\sqrt{E/\rho})$ | $E$ = elastic modulus |
| Rod | Torsion | $L/(\pi\sqrt{G/\rho})$ | $G$ = shear modulus |
| Beam | Bending | $L/(2\pi)(\omega\sqrt{EI/\rho A})^{-1/2}$ | $EI$ = flexural stiffness |
| Membrane | Lateral | $A_s\omega/(2\pi)(N/\rho h)$ | $N$ = stress resultant $A_s$ = surface area |
| Plate | Bending | $A_s/(4\pi)\sqrt{D/\rho h}$ | $D$ = plate flexural stiffness $h$ = plate thickness |
| Acoustic | Dilational | $V_o\omega^2/(2\pi^2)(\sqrt{B/\rho})^3$ | $B$ = bulk modulus $V_o$ = enclosed volume |

structures. The geometry, properties, and boundary conditions for the present example are as follows.

- Geometry: diameter ($D = 300$ in), length ($L = 600$ in), wall thickness ($h = 1$ in).

- Material: aluminum ($E = 10^7$ psi), Poisson's ratio ($\nu = 0.3$), density ($\rho_s g = 0.1$ lb/in$^3$).

- Fluid: water ($\rho_f g = 62.4$ lb/ft$^3 = .036$ lb/in$^3$).

- Fluid B.Cs: fluid free surface at $Z = 600$ in, blocked tank bottom at $Z = 0$ in.

- Structure B.Cs: bottom ($Z$) axial restraint, and free to radially expand without bending restraint.

- Ullage Pressure: $P_0 = 0$ psi, or 30 psi.

The structural shell mode shapes associated with empty and fluid-filled shell are of the form:

$$\Phi(z, \theta; m, n) = \cos(m\pi z/2L) \cdot \cos(n\theta) \quad \text{or} \quad \cos(m\pi z/2L) \cdot \sin(n\theta),$$
$$m = 1, 3, 5, \ldots, n = 0, 1, 2, 3, \ldots. \tag{2.51}$$

Closed-form expressions for empty structure natural frequencies (without and with ullage pressure) are provided below. They are associated with bulge ($n = 0$), lateral ($n - 1$) and shell breathing ($n > 1$) for the empty shell are

$$f_{mn} = \left(\frac{1}{2\pi}\right) \cdot \left(\frac{k_1 + k_2 + k_3}{\rho_s h}\right)^{1/2}, \tag{2.52}$$

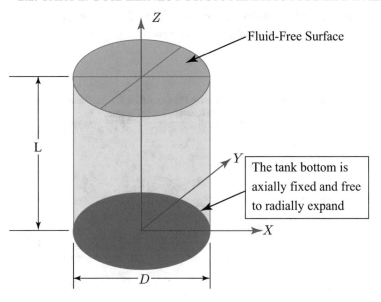

Figure 2.6: Fluid-filled circular cylindrical shell.

where

$$k_1 = \left(\frac{Eh}{R^2}\right) \cdot \frac{\left(\frac{m\pi}{L}\right)^4}{\left(\left(\frac{m\pi}{L}\right)^2 + \left(\frac{n}{R}\right)^2\right)^2} \qquad \text{(membrane stiffness parameter)} \qquad (2.53)$$

$$k_2 = \left(\frac{Eh^3}{12\,(1-v^2)}\right) * \left(\left(\frac{m\pi}{L}\right)^2 + \left(\frac{n}{R}\right)^2\right)^2 \qquad \text{(flexural stiffness parameter)} \qquad (2.54)$$

$$k_3 = (P_o R) \cdot \left(\left(\frac{m\pi}{L}\right)^2 + \left(\frac{n}{R}\right)^2\right) \qquad \text{(ullage "differential" stiffness parameter).} \qquad (2.55)$$

In addition, the empty structure's axial and torsion modal frequencies are (noting that $G = E/(1 + 2v)$),

$$f_{Z,m} = \left(\frac{1}{2\pi}\right) \cdot \left(\frac{m\pi}{L}\right) \cdot \sqrt{E/\rho_s}, \qquad f_{\theta,m} = \left(\frac{1}{2\pi}\right) \cdot \left(\frac{m\pi}{L}\right) \cdot \sqrt{G/\rho_s}, \qquad (2.56)$$

respectively.

Closed-form expressions for the fluid-filled structure natural frequencies (without and with ullage pressure) are presented by Abramson in [19], for the "curious" reader. However, the intent of this illustrative example is to provide insight into parameters affecting the system's modal frequencies.

The first informative result of closed-form modal analysis, provided in Figure 2.7, indicates the roles of membrane, flexural, and ullage pressure strain energies on empty shell natural ($m = 1$) frequencies.

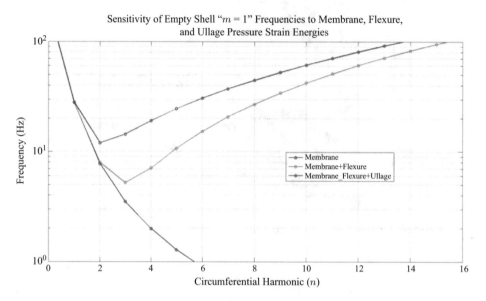

Figure 2.7: Sensitivity of empty shell natural frequencies to strain energy contributions.

The second informative result of closed-form modal analysis, provided in Figure 2.8, indicates the roles of membrane, flexural, and ullage pressure strain energies on fluid-filled shell natural ($m = 1$) frequencies.

The profound effect of fluid mass on shell natural frequencies is summarized in Figure 2.9 for the "no-ullage pressure," $m = 1$ modes.

The high modal density and ullage pressure sensitivity of empty shell natural frequencies are illustrated in Figure 2.10.

Table 2.4 details the numerical natural frequencies corresponding to Figure 2.13. Note that (1) modal frequencies exceeding 85 Hz are shaded in gray and (2) axial and torsion natural frequencies are indicated.

The corresponding high modal density and ullage pressure sensitivity of fluid-filled shell natural frequencies are illustrated in Figure 2.11.

Table 2.5 details the numerical natural frequencies corresponding to Figure 2.11. Note that (1) modal frequencies exceeding 85 Hz are shaded in gray and (2) axial and torsion natural frequencies are indicated.

The following insights are gained from the previous results:

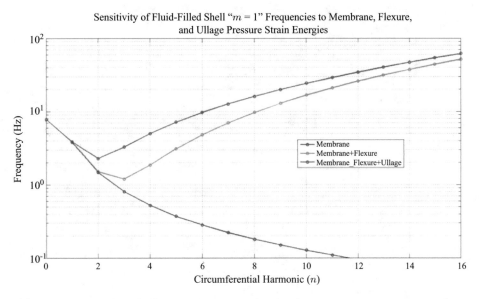

Figure 2.8: Sensitivity of fluid-filled shell natural frequencies to strain energy contributions.

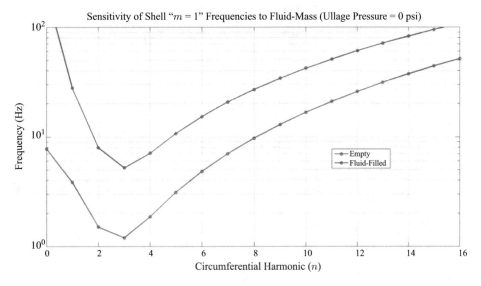

Figure 2.9: Sensitivity of natural frequencies to fluid mass loading.

Table 2.4: Modal density and sensitivity of empty shell natural frequencies to ullage pressure

| | | | Circumferential Harmonic ($n$), $P_0 = 0$ psi | | | | | | | | | | | | | | | | | |
|---|---|---|---|---|---|---|---|---|---|---|---|---|---|---|---|---|---|---|---|---|
| $m$ | Axial | Torsion | 0 (Bulge) | 1 | 2 | 3 | 4 | 5 | 6 | 7 | 8 | 9 | 10 | 11 | 12 | 13 | 14 | 15 | 16 |
| 1 | 81.86 | 64.72 | 208.46 | 27.86 | 7.93 | 5.21 | 7.08 | 10.66 | 15.23 | 20.68 | 26.98 | 34.13 | 42.12 | 50.95 | 60.62 | 71.14 | 82.49 | 94.69 | 107.73 |
| 3 | 245.59 | 194.15 | 208.46 | 121.17 | 53.75 | 28.19 | 18.18 | 15.60 | 17.52 | 21.95 | 27.85 | 34.83 | 42.73 | 51.52 | 61.18 | 71.68 | 83.02 | 95.22 | 108.25 |
| 5 | 409.31 | 323.59 | 208.47 | 165.54 | 102.36 | 62.75 | 41.33 | 30.38 | 26.22 | 26.93 | 30.90 | 36.92 | 44.36 | 52.90 | 62.42 | 72.84 | 84.15 | 96.31 | 109.33 |
| 7 | 573.03 | 453.02 | 208.48 | 184.13 | 136.39 | 95.40 | 67.60 | 50.28 | 40.54 | 36.63 | 37.29 | 41.27 | 47.54 | 55.44 | 64.58 | 74.79 | 85.96 | 98.04 | 111.00 |
| 9 | 736.76 | 582.46 | 208.53 | 193.09 | 158.05 | 121.50 | 92.18 | 71.22 | 57.44 | 49.62 | 46.84 | 48.18 | 52.67 | 59.43 | 67.88 | 77.66 | 88.57 | 100.48 | 113.33 |
| 11 | 900.48 | 711.89 | 208.61 | 198.03 | 171.93 | 141.11 | 113.17 | 90.97 | 74.78 | 64.15 | 58.51 | 57.27 | 59.71 | 65.00 | 72.47 | 81.60 | 92.08 | 103.71 | 116.37 |
| 13 | 1064.21 | 841.33 | 208.75 | 201.08 | 181.16 | 155.65 | 130.37 | 108.54 | 91.35 | 78.96 | 71.23 | 67.84 | 68.32 | 72.04 | 78.33 | 86.63 | 96.54 | 107.78 | 120.18 |
| 15 | 1227.93 | 970.76 | 208.97 | 203.18 | 187.62 | 166.54 | 144.26 | 123.74 | 106.54 | 93.31 | 84.22 | 79.22 | 78.05 | 80.28 | 85.36 | 92.73 | 101.96 | 112.71 | 124.76 |
| 17 | 1391.65 | 1100.20 | 209.30 | 204.78 | 192.38 | 174.89 | 155.49 | 136.71 | 120.19 | 106.81 | 96.99 | 90.90 | 88.47 | 89.43 | 93.36 | 99.80 | 108.29 | 118.49 | 130.13 |
| 19 | 1555.38 | 1229.63 | 209.77 | 206.16 | 196.10 | 181.50 | 164.70 | 147.80 | 132.34 | 119.29 | 109.26 | 102.54 | 99.21 | 99.17 | 102.12 | 107.68 | 115.45 | 125.07 | 136.26 |
| 21 | 1719.10 | 1359.07 | 210.41 | 207.48 | 199.20 | 186.92 | 172.42 | 157.37 | 143.15 | 130.77 | 120.89 | 113.93 | 110.06 | 109.29 | 111.45 | 116.25 | 123.33 | 132.38 | 143.11 |
| 23 | 1882.83 | 1488.50 | 211.26 | 208.84 | 201.95 | 191.57 | 179.04 | 165.74 | 152.85 | 141.34 | 131.89 | 124.96 | 120.84 | 119.60 | 121.17 | 125.34 | 131.84 | 140.36 | 150.64 |
| 25 | 2046.55 | 1617.94 | 212.36 | 210.34 | 204.56 | 195.74 | 184.92 | 173.22 | 161.66 | 151.12 | 142.28 | 135.63 | 131.49 | 130.00 | 131.17 | 134.86 | 140.86 | 148.92 | 158.79 |
| 27 | 2210.27 | 1747.38 | 213.75 | 212.06 | 207.17 | 199.65 | 190.31 | 180.06 | 169.79 | 160.27 | 152.15 | 145.93 | 141.96 | 140.42 | 141.36 | 144.71 | 150.33 | 157.99 | 167.49 |
| 29 | 2374.00 | 1876.81 | 215.48 | 214.04 | 209.89 | 203.47 | 195.42 | 186.49 | 177.42 | 168.93 | 161.61 | 155.94 | 152.27 | 150.82 | 151.68 | 154.83 | 160.16 | 167.51 | 176.70 |
| 31 | 2537.72 | 2006.25 | 217.58 | 216.36 | 212.83 | 207.35 | 200.42 | 192.67 | 184.74 | 177.26 | 170.76 | 165.71 | 162.44 | 161.20 | 162.11 | 165.17 | 170.32 | 177.44 | 186.38 |
| 33 | 2701.45 | 2135.68 | 220.10 | 219.07 | 216.07 | 211.39 | 205.45 | 198.76 | 191.89 | 185.38 | 179.71 | 175.32 | 172.54 | 171.60 | 172.64 | 175.71 | 180.77 | 187.74 | 196.48 |
| 35 | 2865.17 | 2265.12 | 223.09 | 222.21 | 219.67 | 215.69 | 210.62 | 204.90 | 199.01 | 193.42 | 188.57 | 184.86 | 182.60 | 182.03 | 183.29 | 186.45 | 191.50 | 198.38 | 206.99 |
| | | | Circumferential Harmonic ($n$), $P_0 = 30$ psi | | | | | | | | | | | | | | | | |
| $m$ | Axial | Torsion | 0 (Bulge) | 1 | 2 | 3 | 4 | 5 | 6 | 7 | 8 | 9 | 10 | 11 | 12 | 13 | 14 | 15 | 16 |
| 1 | 81.86 | 64.72 | 208.46 | 28.23 | 11.94 | 14.31 | 19.09 | 24.57 | 30.62 | 37.25 | 44.51 | 52.44 | 61.08 | 70.45 | 80.58 | 91.47 | 103.15 | 115.62 | 128.89 |
| 3 | 245.59 | 194.15 | 208.49 | 121.30 | 54.59 | 31.37 | 25.63 | 27.31 | 32.01 | 38.13 | 45.18 | 53.01 | 61.61 | 70.95 | 81.07 | 91.96 | 103.63 | 116.10 | 129.38 |
| 5 | 409.31 | 323.59 | 208.56 | 165.71 | 102.93 | 64.43 | 45.37 | 38.07 | 37.81 | 41.49 | 47.37 | 54.63 | 62.93 | 72.13 | 82.16 | 93.00 | 104.65 | 117.10 | 130.37 |
| 7 | 573.03 | 453.02 | 208.66 | 184.39 | 136.95 | 96.70 | 70.40 | 55.60 | 49.21 | 48.72 | 52.11 | 57.98 | 65.50 | 74.25 | 84.03 | 94.72 | 106.28 | 118.68 | 131.91 |
| 9 | 736.76 | 582.46 | 208.82 | 193.46 | 158.68 | 122.72 | 94.51 | 75.39 | 64.23 | 59.52 | 59.73 | 63.46 | 69.65 | 77.59 | 86.87 | 97.25 | 108.62 | 120.90 | 134.06 |
| 11 | 900.48 | 711.89 | 209.04 | 198.54 | 172.68 | 142.38 | 115.34 | 94.58 | 80.49 | 72.49 | 69.69 | 71.04 | 75.52 | 82.30 | 90.83 | 100.72 | 111.77 | 123.85 | 136.86 |
| 13 | 1064.21 | 841.33 | 209.36 | 201.76 | 182.08 | 157.03 | 132.53 | 111.92 | 96.45 | 86.30 | 81.12 | 80.26 | 82.94 | 88.38 | 95.95 | 105.19 | 115.79 | 127.56 | 140.37 |
| 15 | 1227.93 | 970.76 | 209.78 | 204.06 | 188.73 | 168.08 | 146.50 | 127.04 | 111.33 | 100.02 | 93.19 | 90.55 | 91.58 | 95.66 | 102.18 | 110.65 | 120.70 | 132.07 | 144.60 |
| 17 | 1391.65 | 1100.20 | 210.34 | 205.89 | 193.71 | 176.63 | 157.88 | 140.05 | 124.84 | 113.14 | 105.33 | 101.40 | 101.08 | 103.92 | 109.40 | 117.05 | 126.47 | 137.39 | 149.58 |
| 19 | 1555.38 | 1229.63 | 211.06 | 207.53 | 197.68 | 183.47 | 167.29 | 151.26 | 136.97 | 125.43 | 117.19 | 112.44 | 111.10 | 112.90 | 117.43 | 124.28 | 133.06 | 143.48 | 155.30 |
| 21 | 1719.10 | 1359.07 | 211.99 | 209.12 | 201.05 | 189.16 | 175.23 | 160.99 | 147.86 | 136.84 | 128.58 | 123.40 | 121.38 | 122.37 | 126.10 | 132.22 | 140.39 | 150.30 | 161.72 |
| 23 | 1882.83 | 1488.50 | 213.14 | 210.79 | 204.11 | 194.10 | 182.12 | 169.57 | 157.69 | 147.42 | 139.44 | 134.16 | 131.74 | 132.17 | 135.27 | 140.76 | 148.36 | 157.79 | 168.81 |
| 25 | 2046.55 | 1617.94 | 214.57 | 212.62 | 207.04 | 198.58 | 188.29 | 177.30 | 166.68 | 157.28 | 149.79 | 144.64 | 142.08 | 142.16 | 144.79 | 149.78 | 156.90 | 165.89 | 176.53 |
| 27 | 2210.27 | 1747.38 | 216.31 | 214.68 | 209.99 | 202.82 | 193.99 | 184.42 | 175.02 | 166.56 | 159.69 | 154.86 | 152.34 | 152.26 | 154.59 | 159.20 | 165.92 | 174.53 | 184.82 |
| 29 | 2374.00 | 1876.81 | 218.40 | 217.03 | 213.08 | 206.99 | 199.42 | 191.14 | 182.90 | 175.40 | 169.23 | 164.83 | 162.51 | 162.42 | 164.59 | 168.95 | 175.36 | 183.65 | 193.64 |
| 31 | 2537.72 | 2006.25 | 220.88 | 219.73 | 216.39 | 211.23 | 204.77 | 197.63 | 190.48 | 183.92 | 178.49 | 174.62 | 172.61 | 172.63 | 174.77 | 178.98 | 185.18 | 193.22 | 202.96 |
| 33 | 2701.45 | 2135.68 | 223.80 | 222.83 | 220.01 | 215.64 | 210.15 | 204.05 | 197.91 | 192.26 | 187.59 | 184.29 | 182.66 | 182.90 | 185.10 | 189.26 | 195.33 | 203.19 | 212.73 |
| 35 | 2865.17 | 2265.12 | 227.19 | 226.37 | 224.01 | 220.33 | 215.69 | 210.52 | 205.32 | 200.54 | 196.61 | 193.91 | 192.73 | 193.25 | 195.60 | 199.79 | 205.79 | 213.54 | 222.93 |

Table 2.5: Modal density and sensitivity of fluid-filled shell natural frequencies to ullage pressure

| | | | Circumferential Harmonic ($n$), $P_0 = 0$ psi | | | | | | | | | | | | | | | | |
|---|---|---|---|---|---|---|---|---|---|---|---|---|---|---|---|---|---|---|---|
| $m$ | Axial | Torsion | 0 (Bulge) | 1 (Lateral) | 2 | 3 | 4 | 5 | 6 | 7 | 8 | 9 | 10 | 11 | 12 | 13 | 14 | 15 | 16 |
| 1 | 81.86 | 64.72 | 7.79 | 3.82 | 1.51 | 1.20 | 1.86 | 3.10 | 4.81 | 7.00 | 9.68 | 12.89 | 16.63 | 20.94 | 25.82 | 31.30 | 37.39 | 44.10 | 51.45 |
| 3 | 245.59 | 194.15 | 21.75 | 18.75 | 10.68 | 6.63 | 4.84 | 4.58 | 5.57 | 7.47 | 10.03 | 13.19 | 16.91 | 21.22 | 26.10 | 31.58 | 37.68 | 44.39 | 51.74 |
| 5 | 409.31 | 323.59 | 32.60 | 29.99 | 21.89 | 15.38 | 11.30 | 9.08 | 8.45 | 9.25 | 11.21 | 14.06 | 17.64 | 21.86 | 26.71 | 32.18 | 38.27 | 44.99 | 52.35 |
| 7 | 573.03 | 453.02 | 40.92 | 38.54 | 31.74 | 24.63 | 19.12 | 15.40 | 13.29 | 12.74 | 13.67 | 15.85 | 19.03 | 23.04 | 27.77 | 33.17 | 39.23 | 45.93 | 53.29 |
| 9 | 736.76 | 582.46 | 47.61 | 45.56 | 39.94 | 33.19 | 27.13 | 22.44 | 19.24 | 17.55 | 17.40 | 18.70 | 21.27 | 24.88 | 29.37 | 34.63 | 40.60 | 47.26 | 54.59 |
| 11 | 900.48 | 711.89 | 53.27 | 51.55 | 46.88 | 40.78 | 34.74 | 29.59 | 25.67 | 23.13 | 22.07 | 22.51 | 24.36 | 27.45 | 31.58 | 36.61 | 42.44 | 49.01 | 56.29 |
| 13 | 1064.21 | 841.33 | 58.24 | 56.81 | 52.88 | 47.48 | 41.75 | 36.50 | 32.19 | 29.08 | 27.33 | 27.04 | 28.21 | 30.73 | 34.43 | 39.15 | 44.78 | 51.21 | 58.41 |
| 15 | 1227.93 | 970.76 | 62.73 | 61.53 | 58.19 | 53.44 | 48.14 | 43.02 | 38.58 | 35.14 | 32.92 | 32.07 | 32.65 | 34.62 | 37.87 | 42.25 | 47.63 | 53.89 | 60.97 |
| 17 | 1391.65 | 1100.20 | 66.85 | 65.83 | 62.97 | 58.80 | 53.97 | 49.12 | 44.73 | 41.16 | 38.65 | 37.40 | 37.52 | 39.02 | 41.85 | 45.88 | 50.98 | 57.04 | 63.98 |
| 19 | 1555.38 | 1229.63 | 70.70 | 69.83 | 67.37 | 63.70 | 59.34 | 54.83 | 50.61 | 47.04 | 44.40 | 42.90 | 42.69 | 43.82 | 46.28 | 49.98 | 54.81 | 60.66 | 67.43 |
| 21 | 1719.10 | 1359.07 | 74.35 | 73.61 | 71.47 | 68.24 | 64.33 | 60.19 | 56.22 | 52.77 | 50.12 | 48.50 | 48.07 | 48.92 | 51.08 | 54.50 | 59.07 | 64.71 | 71.32 |
| 23 | 1882.83 | 1488.50 | 77.87 | 77.22 | 75.36 | 72.52 | 69.04 | 65.27 | 61.60 | 58.33 | 55.77 | 54.12 | 53.59 | 54.26 | 56.20 | 59.38 | 63.73 | 69.18 | 75.63 |
| 25 | 2046.55 | 1617.94 | 81.30 | 80.74 | 79.12 | 76.62 | 73.52 | 70.14 | 66.78 | 63.76 | 61.35 | 59.77 | 59.21 | 59.79 | 61.58 | 64.58 | 68.75 | 74.03 | 80.34 |
| 27 | 2210.27 | 1747.38 | 84.70 | 84.21 | 82.80 | 80.61 | 77.87 | 74.85 | 71.83 | 69.08 | 66.87 | 65.42 | 64.91 | 65.46 | 67.17 | 70.06 | 74.10 | 79.24 | 85.43 |
| 29 | 2374.00 | 1876.81 | 88.11 | 87.69 | 86.46 | 84.54 | 82.13 | 79.46 | 76.78 | 74.34 | 72.37 | 71.09 | 70.69 | 71.28 | 72.97 | 75.79 | 79.74 | 84.79 | 90.87 |
| 31 | 2537.72 | 2006.25 | 91.58 | 91.21 | 90.14 | 88.48 | 86.38 | 84.05 | 81.70 | 79.57 | 77.87 | 76.81 | 76.55 | 77.23 | 78.96 | 81.76 | 85.67 | 90.64 | 96.65 |
| 33 | 2701.45 | 2135.68 | 95.15 | 94.83 | 93.91 | 92.48 | 90.67 | 88.66 | 86.64 | 84.82 | 83.41 | 82.58 | 82.51 | 83.32 | 85.12 | 87.96 | 91.85 | 96.80 | 102.76 |
| 35 | 2865.17 | 2265.12 | 98.85 | 98.58 | 97.80 | 96.58 | 95.04 | 93.33 | 91.64 | 90.14 | 89.02 | 88.45 | 88.59 | 89.57 | 91.47 | 94.38 | 98.30 | 103.24 | 109.19 |
| | | | Circumferential Harmonic ($n$), $P_0 = 30$ psi | | | | | | | | | | | | | | | | |
| $m$ | Axial | Torsion | 0 (Bulge) | 1 (Lateral) | 2 | 3 | 4 | 5 | 6 | 7 | 8 | 9 | 10 | 11 | 12 | 13 | 14 | 15 | 16 |
| 1 | 81.86 | 64.72 | 7.79 | 3.87 | 2.27 | 3.29 | 5.02 | 7.15 | 9.68 | 12.61 | 15.97 | 19.80 | 24.12 | 28.95 | 34.32 | 40.25 | 46.75 | 53.85 | 61.55 |
| 3 | 245.59 | 194.15 | 21.75 | 18.77 | 10.84 | 7.38 | 6.83 | 8.02 | 10.18 | 12.97 | 16.27 | 20.08 | 24.38 | 29.22 | 34.59 | 40.52 | 47.03 | 54.13 | 61.84 |
| 5 | 409.31 | 323.59 | 32.62 | 30.03 | 22.01 | 15.79 | 12.40 | 11.38 | 12.18 | 14.24 | 17.19 | 20.81 | 25.03 | 29.81 | 35.16 | 41.09 | 47.60 | 54.70 | 62.42 |
| 7 | 573.03 | 453.02 | 40.96 | 38.59 | 31.86 | 24.96 | 19.92 | 17.02 | 16.13 | 16.95 | 19.11 | 22.27 | 26.22 | 30.86 | 36.13 | 42.01 | 48.50 | 55.60 | 63.33 |
| 9 | 736.76 | 582.46 | 47.68 | 45.64 | 40.10 | 33.52 | 27.82 | 23.76 | 21.52 | 21.06 | 22.19 | 24.63 | 28.13 | 32.48 | 37.58 | 43.36 | 49.79 | 56.87 | 64.57 |
| 11 | 900.48 | 711.89 | 53.38 | 51.69 | 47.09 | 41.15 | 35.41 | 30.77 | 27.63 | 26.14 | 26.29 | 27.92 | 30.81 | 34.76 | 39.58 | 45.19 | 51.52 | 58.52 | 66.20 |
| 13 | 1064.21 | 841.33 | 58.41 | 57.01 | 53.15 | 47.90 | 42.44 | 37.63 | 33.99 | 31.79 | 31.13 | 31.99 | 34.24 | 37.69 | 42.17 | 47.54 | 53.71 | 60.61 | 68.22 |
| 15 | 1227.93 | 970.76 | 62.97 | 61.79 | 58.53 | 53.93 | 48.89 | 44.17 | 40.32 | 37.67 | 36.43 | 36.65 | 38.30 | 41.25 | 45.34 | 50.42 | 56.38 | 63.15 | 70.66 |
| 17 | 1391.65 | 1100.20 | 67.18 | 66.19 | 63.41 | 59.38 | 54.80 | 50.32 | 46.46 | 43.60 | 41.97 | 41.72 | 42.87 | 45.35 | 49.04 | 53.81 | 59.54 | 66.14 | 73.54 |
| 19 | 1555.38 | 1229.63 | 71.14 | 70.29 | 67.91 | 64.39 | 60.27 | 56.11 | 52.38 | 49.46 | 47.63 | 47.05 | 47.80 | 49.88 | 53.21 | 57.68 | 63.17 | 69.58 | 76.85 |
| 21 | 1719.10 | 1359.07 | 74.91 | 74.19 | 72.13 | 69.06 | 65.38 | 61.58 | 58.07 | 55.22 | 53.30 | 52.53 | 53.01 | 54.78 | 57.80 | 61.99 | 67.24 | 73.47 | 80.60 |
| 23 | 1882.83 | 1488.50 | 78.56 | 77.94 | 76.17 | 73.48 | 70.22 | 66.78 | 63.55 | 60.84 | 58.96 | 58.11 | 58.42 | 59.96 | 62.74 | 66.68 | 71.72 | 77.77 | 84.76 |
| 25 | 2046.55 | 1617.94 | 82.15 | 81.61 | 80.08 | 77.73 | 74.86 | 71.79 | 68.85 | 66.36 | 64.59 | 63.74 | 63.98 | 65.38 | 67.97 | 71.72 | 76.58 | 82.47 | 89.32 |
| 27 | 2210.27 | 1747.38 | 85.72 | 85.25 | 83.93 | 81.89 | 79.37 | 76.66 | 74.04 | 71.80 | 70.19 | 69.42 | 69.65 | 70.99 | 73.46 | 77.07 | 81.79 | 87.54 | 94.27 |
| 29 | 2374.00 | 1876.81 | 89.31 | 88.91 | 87.77 | 86.00 | 83.82 | 81.45 | 79.15 | 77.18 | 75.78 | 75.15 | 75.44 | 76.77 | 79.18 | 82.71 | 87.31 | 92.96 | 99.58 |
| 31 | 2537.72 | 2006.25 | 92.97 | 92.63 | 91.65 | 90.14 | 88.26 | 86.22 | 84.24 | 82.56 | 81.40 | 80.94 | 81.34 | 82.71 | 85.12 | 88.60 | 93.14 | 98.70 | 105.25 |
| 33 | 2701.45 | 2135.68 | 96.75 | 96.46 | 95.62 | 94.34 | 92.74 | 91.01 | 89.36 | 87.97 | 87.06 | 86.81 | 87.35 | 88.81 | 91.27 | 94.74 | 99.25 | 104.77 | 111.26 |
| 35 | 2865.17 | 2265.12 | 100.67 | 100.43 | 99.73 | 98.66 | 97.32 | 95.89 | 94.54 | 93.46 | 92.81 | 92.78 | 93.50 | 95.09 | 97.62 | 101.13 | 105.64 | 111.14 | 117.60 |

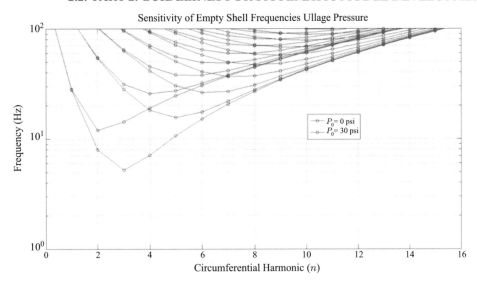

Figure 2.10: Modal density and sensitivity of empty shell natural frequencies to ullage pressure.

a. Bulge ($n = 0$) bulge and lateral ($n = 1$) natural frequencies are *associated with membrane strain energy.*

b. Bulge ($n = 0$) bulge and lateral ($n = 1$) natural frequencies are *insensitive to ullage pressure.*

c. Shell breathing ($n > 1$) natural frequencies are *associated with membrane and flexural strain energies.*

d. Shell breathing ($n > 1$) natural frequencies are *sensitive to ullage pressure.*

e. Shell natural frequencies are *profoundly sensitive to added fluid mass.*

f. Axial and torsion natural frequencies *are unaffected by fluid mass and ullage pressure.*

## 2.2.14    FREE SURFACE SLOSH MODES

A discussion of the modal behavior of a fluid-filled structure is incomplete without inclusion of free surface slosh modes. The pendulum mechanics of slosh are thoroughly discussed by Abramson [19], who described theoretical foundations, experimental results, and a variety of simplified, lumped parameter dynamic models. Fluid-structure interaction FEMs that include free surface effects, developed by Zienkiewicz [8], have been implemented in various finite element software packages. Of primary importance to Abramson's fluid-filled circular cylindrical shell is the fact that the fundamental free surface slosh modes occur at frequencies (0.35 Hz) sufficiently below the lowest flexible structure-fluid mode (1.20 Hz) to be modeled separately as interactions between the sloshing fluid and "rigid" container structure. As a result of situations of this type,

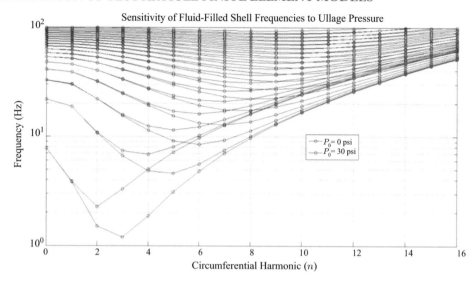

Figure 2.11: Modal density and sensitivity of fluid-filled shell natural frequencies to ullage pressure.

a trivial (zero pressure) boundary condition is often imposed on fluid-filled FEMs. Moreover, since it is quite difficult to instrument the fluid free surface, experimental modal tests most often do not engage in measurement of free surface slosh modes. Inclusion of free surface slosh in predictive models is often accomplished by employing simplified, lumped parameter dynamic models.

## 2.2.15   INTERFACE FLEXIBILITY IN STRUCTURAL ASSEMBLIES—A RETROSPECTIVE

Mathematical solutions for structural components, subjected to a wide variety of boundary conditions are a staple in the historical development of structural mechanics theory [5]. The mathematical solutions are complemented by a wealth of empirical data indicating variability of joint stiffness as well as damping (especially when joints have slip-friction behavior). Significant deviations from assumed ideal joint behavior are also present in structures, which are composed of components that are welded to one another.

A quite revealing illustration of non-ideal boundary conditions is noted in results of a series of modal tests conducted on thin cylindrical shells (see Figure 2.12) at NASA/LARC in the mid-1970s [20].

Initial COSMIC NASTRAN mathematical models of the test article were defined with fixed end boundary conditions for all test conditions: (1) empty, unpressurized; (2) empty pressurized; and (3) half-filled with water unpressurized and pressurized. Natural frequencies of

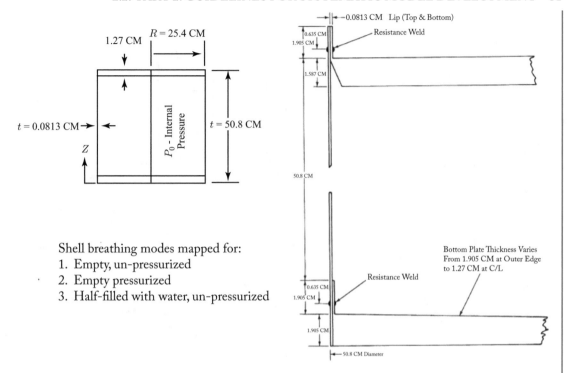

Figure 2.12: **NASA** Langley Research Center cylindrical shell test article.

shell breathing modes for the initial models were significantly higher than all corresponding test data. After changing the model end boundary conditions to pinned (shear diaphragm), which was not intuitively obvious to the young engineer (this book's author), all of the analytical natural frequencies closely followed modal test data, as illustrated in Figure 2.13.

This lesson, experienced by many young engineers, is a clear example of non-ideal boundary conditions that exist in real structures. It is most unfortunate that this point is so easily missed by many practicing engineers due to the high degree of automation in day-to-day utilization of today's computer-aided engineering (CAE) tools.

## 2.2.16    DAMPING IN STRUCTURAL ASSEMBLIES

Most engineering organizations employ empirically based values for modal damping ($\zeta_n$), and the modal approach for description of "viscous" damping, when practical, which circumvents difficulties associated with the lack of a theoretical damping matrix.

A variety of artificially constructed mathematical forms for the damping matrix have been defined over the past century. One form that has managed to find its way into most finite element

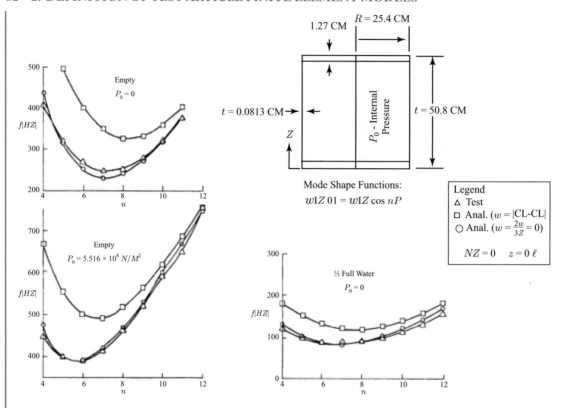

Figure 2.13: Comparison of predicted and measured shell breathing mode frequencies.

codes, namely proportional damping, which is attributed to Rayleigh [21], is

$$[B] = \alpha [M] + \beta [K]. \tag{2.57}$$

Application of the modal transformation on this matrix form results in the following distribution of modal damping, which does not resemble typical empirical data records:

$$\zeta_n = \frac{\alpha}{2\omega_n} + \frac{\beta\omega_n}{2}. \tag{2.58}$$

While proportional damping constructs a mathematically elegant approach for computational structural dynamics, it is never consistent with experimental data, and merits abandonment.

During the late 1920s, Kimball and Lovell [22] and Becker and Foppl [23], independently determined by experiment, that damping in typical structures is simultaneously proportional to displacement (strain) and in phase with velocity, Shortly thereafter, Kussner [24] and Kassner [25] introduced the concept of complex structural damping, which appropriately describes the observations of Kimball, Lovell, Becker, and Foppl. In short, the mathematical description of damping in typical structures shifted from a theoretical (viscous) formulation

$$[M]\{\ddot{u}\} + [B]\{\dot{u}\} + [K]\{u\} = \{F_e(t)\} \qquad (2.59)$$

to a hysteretic formulation (known today as structural damping),

$$[M]\{\ddot{u}\} + (1 + i\eta)[K]\{u\} = \{F_e(t)\}. \qquad (2.60)$$

Note that structural damping and viscous modal damping coefficients are related to one another as

$$\eta = 2\zeta_n. \qquad (2.61)$$

The contributions of Kimball, Lovell, Becker, and Foppl represent a great contrast from the situation which resulted from Rayleigh's citation about proportional damping, which quoted directly from *Theory of Sound* (Chapter V, Paragraph 97) states: "The first case occurs frequently, *in books at any rate*, when motion of each part of the system is resisted by a retarding force, proportional both to the mass and velocity of the part, the same exceptional reduction is possible when *F (the dissipation force)* is a linear function of *T (kinetic energy)* and *V (strain energy)*."

The thorough treatment of structural damping, found in the text by Cremer, Heckl, and Ungar [26], provides a wealth of empirical data along with a technical viewpoint that complements the prevailing, automated finite element mindset. Three crucial features inherent in many structural systems are clearly noted in that text, namely:

1. "Solid" structures generally exhibit damping forces that are independent of frequency (displacement dependent structural damping) for a wide range of building materials.

2. Structural damping is often extremely low ($\eta \sim 10^4$) for individual, unattached structural members, such as bars, beams, plates and shells. This is typical for steel, aluminum and other "hard" metals; damping may be two orders of magnitude greater for lead, concrete, and brick.

3. Structural damping in assemblies is often on the order of $\eta \sim 0.01$, which is attributed to losses in (welded, bolted, riveted, and bearing type) joints.

## 2.2.17 CLOSURE

Practical guidelines for pre-test structural dynamic modeling, presented in this chapter for the integrated test analysis process, are as follows.

1. Determine the relevant frequency band ($0 < f \leq f^*$) by shock and/or response spectrum analysis of anticipated operational environments. In the absence of such data, U.S. government standards (e.g., NASA STD-5002 and SMC-S-004) suggest accepted frequency bands for typical launch vehicle and spacecraft structures.

2. Follow frequency-wavelength guidelines for definition of an appropriate structural dynamic FEM (most modern CAE/FEM software resources by default will inherently satisfy this guideline). However, in many cases engineers take risky "shortcuts" in model definition.

3. The role played by interface flexibility, as noted in the 1970s NASA/LARC shell modal test project, points to the importance of detailed joint modeling as an essential part of FEM modeling. Specifically, models should accurately follow engineering drawings (facilitated by modern CAE/FEM software resources, but often not properly employed) in the vicinity of structural joints.

4. Typical aerospace shell-type structures have relatively high modal density. In particular, overall body bending ($n = 1$), torsional, and shell breathing ($n > 1$) modes occupy the same frequency band. This phenomenon, which occurs for both empty and fluid-filled structures, is descriptive of the *"many modes" problem*. The "many modes" problem, when it is present, significantly affects all steps in the integrated test analysis process.

5. The profound effect of added fluid mass on the fundamental ($m = 1$, $n = 0$) bulge mode of a propellant tank ($\sim$ 208 Hz empty vs. $\sim$ 8 Hz fluid filled for the Abramson example) points to a serious potential deficiency associated with modal testing that is limited empty propellant tank structures. Specifically, a primary structural mode of the flight vehicle is not subjected to verification and validation offered by the integrated test analysis process (typically limited to the 0–50 Hz frequency band).

6. The dominant role of "body" modes in estimation of primary structure flight loads, and (control system and Pogo) stability margins, suggests the potential cost-schedule benefits associated with an integrated test analysis process that focuses primarily on "body" modes ($n = 0$ and 1 for the Abramson example). This consequence offers an opportunity to potentially streamline the modal test plan (Chapter 3).

7. "Body" modes appear to be primarily sensitive to membrane stiffness parameters, and insensitive to shell flexural stiffness parameters and ullage pressure stiffness. This attribute may greatly streamline the test-analysis correlation (Chapter 6) and test-analysis reconciliation (Chapter 7) steps in the integrated test-analysis process.

8. Shell "breathing" modes ($n > 1$ for the Abramson example) are sensitive to membrane, shell flexural, and ullage pressure stiffness parameters. Moreover, these modes are sensitive to local details and manufacturing imperfections. This suggests a significant challenge to the entire integrated test analysis process when "body" and shell breathing modes cannot be separated into distinct classes.

9. In light of widespread misunderstandings related to damping in structures, the assumption of proportional damping should never be employed in structural dynamic modeling. Modal

damping for predicted "real" modes appears to be a safer practice. However, the realities associated with actual damping mechanisms (e.g., concentration at joints and nonlinearity) lead to "unsettling" issues that ultimately point to measured "complex" modal behavior, which cannot be effectively dismissed.

## 2.3  REFERENCES

[1] I. Newton, *Naturalis Principia Mathematica*, London, 1689. 7

[2] J. le Rond d'Alembert, *Trait e de Dynamique*, Paris, 1743. 7

[3] W. R. Hamilton, On a general method in dynamics, *Philosophical Transactions of the Royal Society*, 1835. 7

[4] J. L. Lagrange, *Mechanique Analitique*, Paris, 1788. 7

[5] S. P. Timoshenko, *History of Strength of Materials*, McGraw-Hill, 1953. 7, 30

[6] A. Piersol and T. Paez, Eds., *Harris' Shock and Vibration Handbook*, 6th ed., McGraw-Hill, 2010. 7

[7] W. Ritz, Über eine neue methode zur Lösung gewisser variationsprobleme der mathematischen physik, *Journal Für die Reine und Angewandte Mathematik*, 1909. DOI: 10.1515/crll.1909.135.1. 7, 9

[8] O. C. Zienkiewicz, R. L. Taylor, and J. Z. Zhu, *The Finite Element Method, its Basis and Fundamentals*, 6th ed., Elsevier, 2005. 7, 10, 29

[9] L. A. Pipes, *Matrix Methods for Engineering*, Prentice Hall, 1961. 7, 10

[10] B. G. Galerkin, ...some questions of elastic equilibrium of rods and plates, *Vestnik Inzhenerov i Tekhnikov*, 19, 1915. 7, 10

[11] A. H. Nayfeh and D. T. Mook, *Nonlinear Oscillations*, John Wiley & Sons, 1979. DOI: 10.1002/9783527617586. 7, 10

[12] E. Trefftz, Ein gegenstuck zum ritzschen verfahren, *Proc. of the 2nd International Congress of Applied Mechanics*, 1926. 7, 10

[13] C. A. Brebbia and J. Dominguez, *Boundary Elements an Introductory Course*, 2nd ed., WIT Press, 1992. DOI: 10.1115/1.2897280. 7

[14] W. P. Rodden, *Theoretical and Computational Aeroelasticity*, Crest Publishing, 2011. 8

[15] A. G. Piersol and T. L. Paez, *Harris' Shock and Vibration Handbook*, 6th ed., McGraw-Hill, 2010. 14, 17, 18, 21, 23

[16] S. P. Timoshenko, *History of Strength of Materials*, McGraw-Hill, 1953. 21

[17] Y. C. Fung, *Foundations of Solid Mechanics*, Prentice Hall, 1965. 21

[18] R. H. Lyon and R. G. DeJong, *Theory and Application of Statistical Energy Analysis*, 2nd ed., Butterworth–Heinemann, 1995. 23

[19] H. Norman Abramson, *The Dynamic Behavior of Liquids in Moving Containers*, NASA SP-106, 1966. 23, 25, 29

[20] R. N. Coppolino, *A Numerically Efficient Finite Element Hydroelastic Analysis*, vol. 1 Theory and Results, NASA CR-2662, 1976. DOI: 10.2514/6.1976-1533. 30

[21] L. Rayleigh, *The Theory of Sound*, vol. 1 and vol. 2, Cambridge University Press, 2011. DOI: 10.1063/1.3060230. 32

[22] A. Kimball and D. Lovell, Internal friction in solids, *Physical Review*, 30, 1927. DOI: 10.1103/physrev.30.948. 32

[23] E. Becker and O. Foppl, Dauerversuche zur bestimmung der festigkeitseigenschaften, *Beziehungen Zwischen Baustoffdampfung und Verformungeschwindigkeit*, Forschungsh. Ver. Deutsch. Ing., no. 304, 1928. 32

[24] H. Kussner, Augenblicklicher entwicklungsstand der frage des flugelflatterns, *Luftfahrtforsch*, 12(6):193–209, 1935. 32

[25] R. Kassner, Die berucksichtigung der inneren dampfung beim ebenen problem der flugelschwingung, *Luftfahrtforsch.*, 13(11):388–393, 1936. 32

[26] L. Cremer, M. Heckl, and E. Ungar, *Structure Borne Sound*, Springer-Verlag, 1973. DOI: 10.1007/978-3-662-10121-6. 33

CHAPTER 3

# Systematic Modal Test Planning

## 3.1 PART 1: UNDERSTANDING MODAL DYNAMIC CHARACTERISTICS

### 3.1.1 INTRODUCTION

Comprehensive understanding of a structure's modal characteristics is realized by quantitative analysis of the kinetic and strain energy distributions associated with its normal modes. Analysis of energy distributions offers in-depth insights that are not necessarily realized by visualization of frozen and animated displays of geometric modal deformations.

The modes of an undamped structural dynamic system are solutions of the real, symmetric eigenvalue problem,

$$[K]\{\Phi_n\} - [M]\{\Phi_n\}\lambda_n = \{0\}, \qquad \left(\lambda_n = \omega_n^2\right), \tag{3.1}$$

where $\{\Phi_n\}$ are distinct, orthogonal individual eigenvectors (or mode shapes), and $\lambda_n = \omega_n^2$, are the corresponding eigenvalues. The collection of all (or a truncated set of) eigenvectors defines the system's modal matrix, $[\Phi]$, which (when normalized to unit modal mass) has the following properties:

$$[\Phi]^T [M] [\Phi] = [I], \qquad [\Phi]^T [K] [\Phi] = [\lambda]. \tag{3.2}$$

An interesting aspect of modal orthogonality for an undamped structural dynamic system lies in the fact that the orthogonality relationships can be "unpacked" to describe modal kinetic and strain energy distributions. Specifically, the respective kinetic and strain distributions for each mode are the term-by term products:

$$\{KE\}_n = \{[M]\{\Phi\}_n\} \otimes \{\Phi\}_n, \qquad KE_{TOT,n} = \sum_{i=1}^{DOF} KE_{in} = 1$$

$$\{SE\}_n = \{[K]\{\Phi\}_n\} \otimes \{\Phi\}_n /\lambda_n, \qquad SE_{TOT,n} = \sum_{i=1}^{DOF} SE_{in} = 1. \tag{3.3}$$

The individual terms in each of these "energy" vectors are directly associated with the dynamic system degrees of freedom. As such, they provide appropriately weighted metrics for

kinetic and strain energy distributions that are not indicated by the geometric modes shape (e.g., "heavier" degrees of freedom have greater kinetic energy than "lighter" degrees of freedom with equivalent modal displacements). A direct consequence of the algebraic eigenvalue problem, Equation (3.1), is that the modal kinetic and strain energy distributions for the assembled system are identical. However, in a later discussion, segmentation of the system's mass and stiffness distributions into component partitions will be demonstrated as a means of discriminating distributions modal kinetic and strain energies. That being said, modal kinetic energy grouped sums, partitioned by global direction yield the direction of overall modal activity for a particular mode, depicted as follows:

$$
\begin{aligned}
\{KE\}_n \Rightarrow & \sum \{KE_{TX}\}_n, \quad \sum \{KE_{TY}\}_n, \quad \sum \{KE_{TZ}\}_n, \\
& \quad \sum \{KE_{RX}\}_n, \quad \sum \{KE_{RY}\}_n, \quad \sum \{KE_{RZ}\}_n \\
\sum \{KE_{TX}\}_n + & \sum \{KE_{TY}\}_n + \sum \{KE_{TZ}\}_n \\
& + \sum \{KE_{RX}\}_n + \sum \{KE_{RY}\}_n + \sum \{KE_{RZ}\}_n = 1.
\end{aligned}
\tag{3.4}
$$

In addition, modal kinetic energy grouped sums, partitioned by subsystem component, provide additional valuable insights, i.e.,

$$
\begin{aligned}
\{KE\}_n \Rightarrow & \sum \{KE_1\}_n, \quad \sum \{KE_2\}_n, \quad \sum \{KE_3\}_n, \dots \\
& \sum \{KE_1\}_n + \sum \{KE_2\}_n + \sum \{KE_3\}_n, \dots = 1.
\end{aligned}
\tag{3.5}
$$

### 3.1.2   ILLUSTRATIVE EXAMPLE: ISS P5 SHORT SPACER

The ISS (International Space Station) P5 Short Spacer was the subject an end-to-end exercise of the Integrated Test Analysis Process in 2001. Modal testing was conducted by a team composed of Boeing/Rocketdyne (the manufacturer), NASA/MSFC (the laboratory), and Measurement Analysis Corporation (the consultant). The P5 test article and (space shuttle payload bay geometric) laboratory test fixture are shown in Figure 3.1.

The 21,666 DOF (3611 grid points) FEM of the ISS P5 (weighing 3605 lb) and test fixture is illustrated in Figure 3.2.

Descriptive geometric plots of the first two FEM modes, illustrated in Figure 3.3, reveal the character and content of modal behavior in a superficial manner.

A summary of modal kinetic energies associated with the first 36 FEM modes of the P5 test article (note that "rotational DOF distributions are not indicated as they were negligible) is provided in Table 3.1.

Note that the first 13 FEM modes have less than 1% fixture modal kinetic energy, and non-negligible fixture modal kinetic energy occurs in most of the higher FEM modes. In addition, modes with dominant fixture modal kinetic energy occur at 50 Hz and higher frequencies.

Figure 3.1: ISS P5 short spacer.

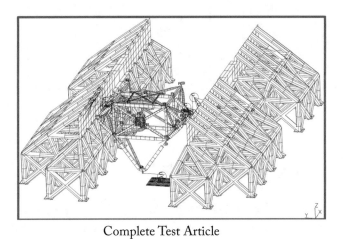

Complete Test Article

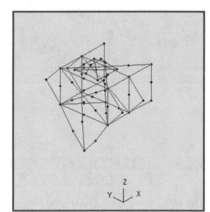

P5 Truss (red), P5 Grapple (blue)

Figure 3.2: ISS P5 and test fixture finite element model.

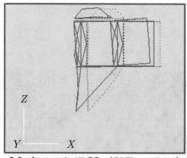
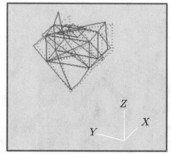

Mode 1: 18.67 Hz ($KE_X$ = 94%)     Mode 2: 19.31 Hz ($KE_Y$ = 89%)

Figure 3.3: Mode 1: 18.67 Hz ($KE_x = 94\%$), Mode 2: 19.31 Hz ($KE_y = 89\%$).

Table 3.1: P5 test article FEM modal kinetic energy distributions

| Mode | Freq (Hz) | Directional $KE$ (%) | | | Component $KE$ (%) | | | Mode | Freq (Hz) | Directional $KE$ (%) | | | Component $KE$ (%) | | |
|---|---|---|---|---|---|---|---|---|---|---|---|---|---|---|---|
| | | $X$ | $Y$ | $Z$ | Truss | Grapple | Fixture | | | $X$ | $Y$ | $Z$ | Truss | Grapple | Fixture |
| 1 | 18.67 | 94 | 2 | 4 | 56 | 44 | 0 | 19 | 52.17 | 15 | 41 | 43 | 74 | 4 | 22 |
| 2 | 19.31 | 10 | 89 | 1 | 93 | 7 | 0 | 20 | 52.57 | 3 | 59 | 38 | 68 | 1 | 31 |
| 3 | 24.88 | 84 | 4 | 12 | 56 | 44 | 0 | 21 | 52.8 | 2 | 87 | 11 | 3 | 0 | 97 |
| 4 | 28.68 | 46 | 15 | 39 | 75 | 25 | 0 | 22 | 53.11 | 14 | 58 | 28 | 46 | 3 | 52 |
| 5 | 29.37 | 35 | 24 | 41 | 63 | 37 | 0 | 23 | 54.41 | 49 | 25 | 27 | 97 | 1 | 2 |
| 6 | 30.07 | 14 | 23 | 63 | 39 | 61 | 0 | 24 | 55.9 | 44 | 15 | 41 | 98 | 2 | 0 |
| 7 | 34.01 | 49 | 31 | 20 | 96 | 4 | 0 | 25 | 56.96 | 21 | 7 | 72 | 76 | 23 | 0 |
| 8 | 34.65 | 12 | 23 | 65 | 99 | 1 | 0 | 26 | 58.03 | 79 | 19 | 2 | 99 | 1 | 0 |
| 9 | 35.22 | 42 | 51 | 7 | 77 | 23 | 0 | 27 | 59.83 | 0 | 95 | 5 | 2 | 0 | 98 |
| 10 | 35.61 | 40 | 25 | 36 | 89 | 11 | 0 | 28 | 59.86 | 1 | 96 | 4 | 1 | 0 | 99 |
| 11 | 37.82 | 27 | 51 | 22 | 56 | 44 | 0 | 29 | 61.23 | 70 | 5 | 26 | 98 | 2 | 1 |
| 12 | 40.82 | 7 | 18 | 75 | 32 | 68 | 0 | 30 | 63.2 | 0 | 2 | 98 | 2 | 98 | 0 |
| 13 | 42.24 | 6 | 2 | 93 | 16 | 84 | 0 | 31 | 64.76 | 98 | 1 | 1 | 0 | 0 | 100 |
| 14 | 45.21 | 22 | 11 | 67 | 91 | 7 | 2 | 32 | 64.76 | 98 | 1 | 1 | 0 | 0 | 100 |
| 15 | 47.4 | 26 | 52 | 22 | 95 | 4 | 1 | 33 | 65.95 | 3 | 9 | 88 | 68 | 27 | 4 |
| 16 | 50.22 | 6 | 67 | 27 | 19 | 4 | 76 | 34 | 68.66 | 3 | 77 | 20 | 3 | 0 | 97 |
| 17 | 50.41 | 10 | 77 | 13 | 14 | 0 | 86 | 35 | 68.74 | 2 | 72 | 25 | 7 | 2 | 91 |
| 18 | 50.95 | 66 | 23 | 11 | 87 | 0 | 13 | 36 | 69.47 | 76 | 17 | 7 | 98 | 1 | 1 |

### 3.1.3  ILLUSTRATIVE EXAMPLE: AXISYMMETRIC SHELL FINITE ELEMENT MODEL

The shell structure, shown in Figure 3.4, consists of five substructures, namely: (1) a lower cylindrical skirt (fully fixed at its base), (2) a lower hemispherical bulkhead, (3) lower cylindrical section, (4) upper cylindrical section, and (5) upper hemispherical bulkhead.

In addition, allocation of unassembled component mass and stiffness matrices within the 5616 DOF model are illustrated in Figure 3.5.

The overall dimensions of the aluminum structure are length, $L = 100$ in, radius, $R = 20$ in, and wall thickness, $h = 0.4$ in. It should be noted that this illustrative example structure does not represent a realistic design. The rather high thickness-to-radius ratio, $h/R = 1/50$, was

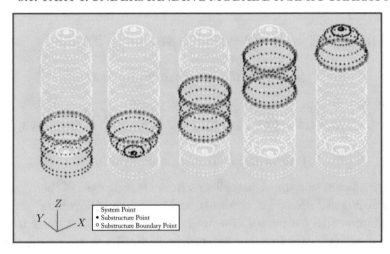

Figure 3.4: Illustrative example shell structure.

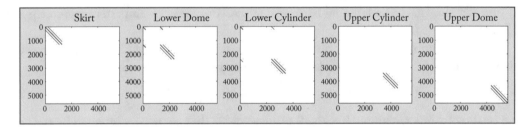

Figure 3.5: Allocation of shell structure component matrices.

selected to produce less shell breathing modes in the base frequency band ($f < 2000$ Hz) than typical aerospace systems, while including modes of sufficient complexity to illustrate key aspects of quantitative normal mode metrics. The subject structure, fully constrained at the bottom of the lower cylindrical skirt, has 150 modes in the 0–1453 Hz frequency band. If the example structure is viewed as a 1/20th scale model, the full-scale set of 150 modal frequencies becomes 0–73 Hz, which is representative of the range of interest for some spacecraft applications.

Modal kinetic and strain energy grouped sums, partitioned by subsystem component, yield the distribution of component activity for a particular mode. It should be noted that the component kinetic and strain energies are not necessarily distributed in the same manner as one another when they are computed on the basis of separate component mass and stiffness matrices in accordance with

$${KE_m}_n = {[M_m]{\Phi}_n} \otimes {\Phi}_n, \quad KE_{mn} = \sum_{DOF} {KE_m}_n \leq 1$$

(modal $KE$ for component $m$, mode $n$)

$$(3.6)$$

$${SE_m}_n = {[K_m]{\Phi}_n} \otimes {\Phi}_n / \lambda_n, \quad SE_{mn} = \sum_{DOF} {SE_m}_n \leq 1$$

(modal $SE$ for component $m$, mode $n$).

Modal dynamics of axisymmetric shell structures are composed of "body" modes (i.e., axial, lateral two orthogonal directions, torsional, and bulge) and "breathing" modes associated with $n > 1$ circumferential harmonics as well as axial harmonics). In order to systematically classify modes of geometrically (but not necessarily physically) axisymmetric shell structures, consider the seven "body" geometric patterns for successive shell circumferential stations illustrated in Figure 3.6.

The seven patterns are associated with cross-sectional lateral ("$TX$" and "$TY$"), and axial ("$TZ$") translations, pitch, yaw, and torsional ("$RX$,""$RY$," and "$RZ$") rotations, and radial bulge ("$TR$") translation. In "circumferential harmonic terms," the above seven patterns represent n=0 and 1 motions (or load patterns).

By organizing the above-described geometric patterns as a body displacement transformation matrix, $[\psi_b]$, the discrete FEM shell displacements, $[\Phi]$, are expressed as

$$[\Phi] = [\Phi_b] + [\Phi_r] = [\psi_b][\varphi_b] + [\Phi_r],$$

$$(3.7)$$

where $[\Phi_r]$ represents remaining (residual) displacements that are not represented by the body patterns.

Employing weighted linear least-squares analysis, as described below, each system normal mode may be partitioned into "body" and (remainder) "breathing" components, as follows.

Step 1: Weighted least squares

$$[\Psi_b^T M \Phi] = [\Psi_b^T M \Psi_b][\varphi_b] + [\Psi_b^T M \Phi_r].$$

$$(3.8)$$

Step 2: Orthogonal constraint

$$[\psi_b^T M \Phi_r] = [0].$$

$$(3.9)$$

Step 3: Generalized "body" modes

$$[\varphi_b] = [\Psi_b^T M \Psi_b]^{-1}[\Psi_b^T M \Phi].$$

$$(3.10)$$

Step 4: "Body" and "breathing" distributions

$$[KE_b] = [M\Phi_b] \otimes [\Phi_b], \quad [KE_r] = [M\Phi_r] \otimes [\Phi_r].$$

$$(3.11)$$

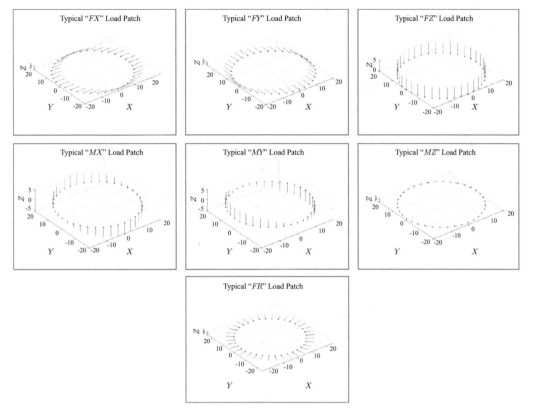

Figure 3.6: Shell axial station "body" geometric deformation patterns.

Step 5: "Body" and "breathing" total kinetic energies formed by column sums of Equation (3.11):

$$\text{Body kinetic energy for mode "}n\text{"} = KE_{b,n} = \sum \{KE_b\}_n$$
$$\text{Breathing kinetic energy for mode "}n\text{"} = KE_{r,n} = \sum \{KE_r\}_n \,.$$

(3.12)

The sum of "body" and "breathing" modal kinetic energies for each mode is always unity (or 100%). In the case of an ideal axisymmetric structure, the "body" and "breathing" modal kinetic energy distributions are either 0 or 100%. When imperfections and/or localized features are present (defining a perturbed shell structure) designations of "body" and "breathing" modal kinetic energy distributions are less distinct. In such situations, the modal kinetic energy designation of mode "type" may be defined on the basis of majority components (i.e., a "body" mode has more than 50% body kinetic energy).

A yet more subtle description of "body" and "breathing" modal kinetic energies is realized by analysis of the "body" contribution in Equation (3.7), specifically since

$$[\Phi_b] = [\Psi_b][\varphi_b],$$
(3.13)

$$\left[\Phi_b^T M \Phi_b\right] = \left[\varphi_b^T\right]\left[\Psi_b^T M \Psi_b\right][\varphi_b] = \left[\varphi_b^T\right][m_b][\varphi_b].$$
(3.14)

Therefore, the unpacking of Equation (3.14) results in the "body" mode kinetic energy distribution expressed in terms of generalized coordinates associated with the seven shape patterns per axial station, i.e.,

$$[KE_b] = [m\varphi_b] \otimes [\varphi_b].$$
(3.15)

This result offers the opportunity to describe the "body" modal kinetic energy distributions in terms of generalized "directions." Specifically,

a. "body $X$" bending kinetic energy:

$$KE_{bx,n} = \sum_{1,8,\ldots} \{KE_b\}_n + \sum_{5,12,\ldots} \{KE_b\}_n.$$
(3.16)

b. "body $Y$" bending kinetic energy:

$$KE_{by,n} = \sum_{2,9,\ldots} \{KE_b\}_n + \sum_{4,11,\ldots} \{KE_b\}_n.$$
(3.17)

c. "body $Z$" axial kinetic energy:

$$KE_{bz,n} = \sum_{3,10,\ldots} \{KE_b\}_n.$$
(3.18)

d. "body" torsion kinetic energy:

$$KE_{bt,n} = \sum_{6,13,\ldots} \{KE_b\}_n.$$
(3.19)

e. "body $(n = 0)$" bulge kinetic energy:

$$KE_{bb,n} = \sum_{7,14,\ldots} \{KE_b\}_n.$$
(3.20)

The sixth "breathing" kinetic energy is described by the previous result in Equation (3.12), specifically

f. "breathing" kinetic energy:

$$KE_{r,n} = \sum_{1,2,3,\ldots} \{KE_r\}_n.$$
(3.21)

A summary of ideal axisymmetric shell and perturbed shell structure modal frequencies and kinetic energy metrics (sorted in terms of total "body" and "breathing" kinetic energy distributions (Equation set (3.13)), for ideal axisymmetric and perturbed shells, is presented in Figure 3.7.

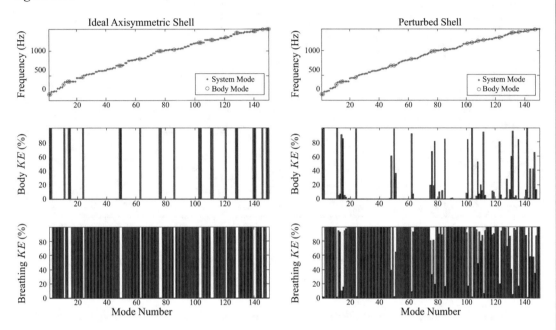

Figure 3.7: Axisymmetric and perturbed shell structure modal frequencies and kinetic energy metrics.

The above result is quite significant, as the vast majority of system modes are associated with "breathing" deformation, which will be shown in the subsequent discussion on Target Mode Selection to be relatively insignificant contributors to primary structure dynamic loads.

An in-depth summary of the present illustrative example's modal metrics, for a representative subset of the first 150 FEM modes for an ideal axisymmetric shell is provided in Table 3.2.

The following observations result from analysis of modal energy characteristics for the ideal axisymmetric structure.

1. Twenty-four "body" modes, 126 ($n > 1$) "breathing" modes.

2. Nine "body" bending modes in each of the "$X$" and "$Y$" directions.

3. Two "body" axial "$Z$" modes.

4. Two "body" torsion modes.

5. Two "body" mixed axial-bulge modes.

Table 3.2: Selected modes of the example shell structure

| Mode | Freq (Hz) | Body & Breathing Kinetic Energies (%) | | Body Mode Kinetic Energy Types (%) | | | | | Component Kinetic Energies (%) | | | | | Component Strain Energies (%) | | | | |
|---|---|---|---|---|---|---|---|---|---|---|---|---|---|---|---|---|---|---|
| | | Body | Breathing | X-Bending | Y-Bending | Z-Axial | Torsion | n=0 Bulge | Skirt | Dome 1 | Shell 1 | Shell 2 | Dome 2 | Skirt | Dome 1 | Shell 1 | Shell 2 | Dome 2 |
| 1 | 122.20 | 100 | | | 100 | | | | 1 | 1 | 11 | 36 | 51 | 65 | 0 | 27 | 8 | 1 |
| 2 | 122.20 | 100 | | 100 | | | | | 1 | 1 | 11 | 36 | 51 | 65 | 0 | 27 | 8 | 1 |
| 3 | 176.09 | | 100 | 100 | | | | | 1 | 1 | 46 | 51 | 1 | 14 | 7 | 35 | 41 | 3 |
| 4 | 176.09 | | 100 | | 100 | | | | 1 | 1 | 46 | 51 | 1 | 14 | 7 | 35 | 41 | 3 |
| 11 | 315.24 | 100 | | | | 100 | | | 4 | 6 | 22 | 43 | 25 | 59 | 0 | 31 | 10 | 0 |
| 12 | 375.44 | | 100 | | | | | | 0 | 0 | 45 | 54 | 0 | 1 | 0 | 44 | 54 | 0 |
| 13 | 375.46 | | 100 | | | | | | 0 | 0 | 45 | 54 | 0 | 1 | 0 | 44 | 54 | 0 |
| 14 | 377.21 | 100 | | | 100 | | | | 12 | 35 | 28 | 9 | 16 | 67 | 5 | 15 | 11 | 2 |
| 15 | 377.21 | 100 | | 100 | | | | | 12 | 35 | 28 | 9 | 16 | 67 | 5 | 15 | 11 | 2 |
| 16 | 377.77 | | 100 | | | | | | 1 | 1 | 51 | 45 | 1 | 4 | 5 | 44 | 44 | 3 |
| 24 | 467.77 | 100 | | | | 100 | | | 3 | 7 | 18 | 35 | 36 | 56 | 1 | 29 | 12 | 3 |
| 25 | 506.44 | | 100 | | | | | | 96 | 1 | 2 | 1 | 0 | 85 | 12 | 1 | 2 | 0 |
| 26 | 506.44 | | 100 | | | | | | 96 | 1 | 2 | 1 | 0 | 85 | 12 | 1 | 2 | 0 |
| 49 | 706.65 | 100 | | | | 100 | | | 3 | 36 | 15 | 27 | 19 | 10 | 21 | 38 | 23 | 9 |
| 50 | 706.65 | 100 | | 100 | | | | | 3 | 36 | 15 | 27 | 19 | 10 | 21 | 38 | 23 | 9 |
| 63 | 841.31 | 100 | | | | | 100 | | 18 | 28 | 22 | 11 | 21 | 34 | 4 | 24 | 35 | 3 |
| 76 | 997.48 | 100 | | | | 100 | | | 17 | 18 | 27 | 25 | 13 | 20 | 13 | 29 | 37 | 2 |
| 77 | 997.50 | 100 | | 100 | | | | | 17 | 18 | 27 | 25 | 13 | 20 | 13 | 29 | 37 | 2 |
| 86 | 1029.94 | 100 | | | | | 96 | 4 | 8 | 50 | 11 | 5 | 27 | 25 | 26 | 12 | 21 | 16 |
| 103 | 1169.56 | 100 | | | | 100 | | | 25 | 17 | 10 | 23 | 24 | 23 | 24 | 7 | 27 | 18 |
| 104 | 1169.59 | 100 | | 100 | | | | | 25 | 17 | 10 | 23 | 24 | 23 | 24 | 7 | 27 | 18 |
| 111 | 1223.46 | 100 | | | | 100 | | | 26 | 2 | 28 | 28 | 16 | 25 | 1 | 29 | 22 | 23 |
| 112 | 1223.49 | 100 | | 100 | | | | | 26 | 2 | 28 | 28 | 16 | 25 | 1 | 29 | 22 | 23 |
| 121 | 1273.43 | 100 | | | | | 81 | 19 | 3 | 33 | 11 | 14 | 39 | 4 | 34 | 11 | 10 | 41 |
| 128 | 1368.34 | 100 | | | | 100 | | | 22 | 4 | 34 | 31 | 8 | 24 | 5 | 33 | 28 | 10 |
| 129 | 1368.41 | 100 | | 100 | | | | | 22 | 4 | 34 | 31 | 8 | 24 | 5 | 33 | 28 | 10 |
| 140 | 1402.79 | 99 | 1 | | | 99 | | | 9 | 44 | 14 | 7 | 26 | 16 | 33 | 14 | 16 | 20 |
| 141 | 1402.83 | 99 | 1 | 99 | | | | | 9 | 44 | 14 | 7 | 26 | 16 | 33 | 14 | 16 | 20 |
| 146 | 1440.38 | 100 | | | | | 45 | 55 | 9 | 15 | 29 | 29 | 17 | 10 | 15 | 29 | 29 | 17 |
| 149 | 1452.66 | 99 | 1 | | | 99 | | | 15 | 18 | 23 | 24 | 19 | 15 | 18 | 24 | 24 | 19 |
| 150 | 1452.71 | 99 | 1 | 99 | | | | | 15 | 18 | 24 | 24 | 19 | 15 | 18 | 24 | 24 | 19 |

6. Contrasting kinetic and strain energy distributions in fundamental bending, axial, torsion modes.

Regarding the sixth observation, the contrasting kinetic and strain energy distributions (i.e., dominant kinetic energy toward the top, dominant strain energy toward the foundation) conform to fundamental mechanical expectations.

Graphical illustrations indicating the character of the fundamental ($Y$) bending mode and the lowest frequency shell breathing mode are provided in Figure 3.8. Kinetic ($KE$) and strain or potential energy ($PE$) distributions are indicated in "pie" format.

## 3.1.4  CLOSURE

Comprehensive understanding of a structure's modal characteristics employing modal kinetic and strain energy metrics has been demonstrated with two key illustrative examples, namely: (a) the International Space Station P5 test article, which has benefitted from an end-to-end integrated test-analysis project and (b) the axisymmetric shell finite element model, which addresses aspects of the "many modes" problem.

Unpacking of fundamental orthogonality relationships mathematically yields modal kinetic and strain energy distributions. Physically meaningful distributions of modal kinetic and strain energy distributions occur only if component mass and stiffness matrices are subdivided into constituent, decoupled partitions; otherwise the assembled system kinetic and strain energy distributions must be identical (a mathematical consequence of the algebraic eigenvalue prob-

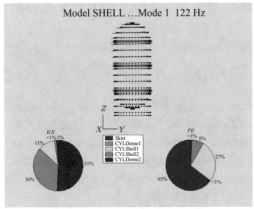

Mode 1: $Y$-Bending                    Mode 3: Cylinder $n = 3$ Breathing

Figure 3.8: Graphical illustration of the character of two system modes.

lem solution). Classification of body mode properties (i.e., bending, axial, torsion, $n = 0$ bulge) is made possible through the utilization of body geometric deformation patterns.

The "many modes" problem is partially managed by classification of a minority of the population as "body" modes. Additional guidance through the "many modes" problem is provided by the process of target mode selection, which is the topic of the next discussion.

## 3.2    PART 2: TARGET MODE SELECTION

### 3.2.1    INTRODUCTION

Guidelines for target mode selection (a subset of "significant" system modes within the frequency band, $0 \leq f \leq f^*$) discussed in Chapter 2, are documented in NASA STD-5002 [1]. While that document appropriately specifies criteria for dynamic systems excited by boundary accelerations, it is not sufficiently clear in its reference to "augmented by modes which are critical for specific load or deflection definition." The present discussion offers further clarification regarding target mode selection for cases in which the system is excited by (a) boundary accelerations and/or (b) applied external loads.

In Chapter 2, matrix equations associated with response of a linear structural dynamic system to dynamic excitations were described. That previous discussion is generalized herein to include the potential presence of localized nonlinearities, primarily at joints, which occur in some aerospace systems.

## 3.2.2   STRUCTURE EXCITED BY BOUNDARY ACCELERATIONS AND APPLIED EXTERNAL LOADS

The form of matrix structural dynamic equations with localized nonlinearities is

$$[M]\{\ddot{U}\} + [B]\{\dot{U}\} + [K]\{U\} = [\Gamma_e]\{F_e\} + [\Gamma_N]\{F_N\}. \tag{3.22}$$

Symmetric system mass, damping, and stiffness matrices, $[M]$, $[B]$, and $[K]$, respectively, are associated with discrete displacement degrees of freedom, $\{U\}$. The external loads, $\{F_e\}$ are allocated to the discrete displacements in accordance with geometric distributions described by the (linearly independent) columns of the matrix, $[\Gamma_e]$. Moreover, localized nonlinear internal loads, $\{F_N\}$ are allocated to the discrete displacements in accordance with geometric distributions described by the (linearly independent) columns of the matrix, $[\Gamma_N]$.

The vital companion displacements associated with the nonlinear internal loads are

$$\{U_N\} = \left[\Gamma_N^T\right]\{U\}. \tag{3.23}$$

Since the most general case nonlinear internal loads may be described as nonlinear functions of localized displacements, velocities, and parameters, the nonlinear loads follow a symbolic algorithm:

$$\{F_N, p\}_{t+\Delta t} = \{f\left(U_N, \dot{U}_N, p\right)\}_t. \tag{3.24}$$

If the parameters ($p$) are updated as a function of time (or motion path), the nonlinearities are "hysteretic." Alternatively, if the parameters ($p$) are fixed, the nonlinearities are "algebraic."

By partitioning the displacements into "interior" and "boundary" subsets,

$$\{U\} = \left\{ \begin{array}{c} U_i \\ U_b \end{array} \right\}. \tag{3.25}$$

The system equations (assuming for the purposes of this discussion that localized nonlinear forces are associated with interior DOFs) are expressed as:

$$\begin{array}{c} \left[ \begin{array}{cc} M_{ii} & M_{ib} \\ M_{bi} & M_{bb} \end{array} \right] \left\{ \begin{array}{c} \ddot{U}_i \\ \ddot{U}_b \end{array} \right\} + \left[ \begin{array}{cc} B_{ii} & B_{ib} \\ B_{bi} & B_{bb} \end{array} \right] \left\{ \begin{array}{c} \dot{U}_i \\ \dot{U}_b \end{array} \right\} + \left[ \begin{array}{cc} K_{ii} & K_{ib} \\ K_{bi} & K_{bb} \end{array} \right] \left\{ \begin{array}{c} U_i \\ U_b \end{array} \right\} \\ = \left[ \begin{array}{cc} \Gamma_{ie} & 0_{ib} \\ 0_{bi} & I_{bb} \end{array} \right] \left\{ \begin{array}{c} F_{ie} \\ F_b \end{array} \right\} + \left[ \begin{array}{c} \Gamma_{iN} \\ 0_{bN} \end{array} \right] \{F_N\}. \end{array} \tag{3.26}$$

Note that the boundary forces, $\{F_b\}$, are not designated as "exterior loads" or "interface loads" with another structure, but simply "boundary loads."

The transformation of "interior" displacements to "relative" displacements with respect to the "boundary" displacements is defined as:

$$\left\{ \begin{array}{c} U_i \\ U_b \end{array} \right\} = \left[ \begin{array}{cc} I_{ii} & -K_{ii}^{-1}K_{ib} \\ 0_{bi} & I_{bb} \end{array} \right] \left\{ \begin{array}{c} U'_i \\ U_b \end{array} \right\} = \left[ \begin{array}{cc} I_{ii} & \Psi_{ib} \\ 0_{bi} & I_{bb} \end{array} \right] \left\{ \begin{array}{c} U'_i \\ U_b \end{array} \right\}. \tag{3.27}$$

Resulting in the transformed system equations,

$$
\begin{bmatrix} M_{ii} & M'_{ib} \\ M'_{bi} & M'_{bb} \end{bmatrix} \begin{Bmatrix} \ddot{U}'_i \\ \ddot{U}_b \end{Bmatrix} + \begin{bmatrix} B_{ii} & B'_{ib} \\ B'_{bi} & B'_{bb} \end{bmatrix} \begin{Bmatrix} \dot{U}'_i \\ \dot{U}_b \end{Bmatrix} + \begin{bmatrix} K_{ii} & 0_{ib} \\ 0_{bi} & K'_{bb} \end{bmatrix} \begin{Bmatrix} U'_i \\ U_b \end{Bmatrix}
$$
$$
= \begin{bmatrix} \Gamma_{ie} & 0_{ib} \\ \Psi_{ib}^T \Gamma_{ie} & I_{bb} \end{bmatrix} \begin{Bmatrix} F_{ie} \\ F_b \end{Bmatrix} + \begin{bmatrix} \Gamma_{iN} \\ 0_{bN} \end{bmatrix} \{F_N\} .
$$

(3.28)

Application of the modal transformation (real, undamped modes) of the upper partition of Equation (3.27),

$$
\begin{Bmatrix} U'_i \\ U_b \end{Bmatrix} = \begin{bmatrix} \Phi_{in} & 0_{ib} \\ 0_{bi} & I_{bb} \end{bmatrix} \begin{Bmatrix} q_n \\ U_b \end{Bmatrix}, \quad [\Phi_{in}]^T [M_{ii}] [\Phi_{in}] = [I_{nn}],
$$
$$
[\Phi_{in}]^T [K_{ii}] [\Phi_{in}] = [\omega_n^2],
$$

(3.29)

results in the "modal" dynamic equations,

$$
\begin{bmatrix} I_{nn} & P_{nb} \\ P_{bn} & M'_{bb} \end{bmatrix} \begin{Bmatrix} \ddot{q}_n \\ \ddot{U}_b \end{Bmatrix} + \begin{bmatrix} 2\xi_n\omega_n & 0_{nb} \\ 0_{bn} & B'_{bb} \end{bmatrix} \begin{Bmatrix} \dot{q}_n \\ \dot{U}_b \end{Bmatrix} + \begin{bmatrix} \omega_n^2 & 0_{nb} \\ 0_{bn} & K'_{bb} \end{bmatrix} \begin{Bmatrix} q_n \\ U_b \end{Bmatrix}
$$
$$
= \begin{bmatrix} \Phi_{in}^T \Gamma_{ie} & 0_{nb} \\ \Psi_{ib}^T \Gamma_{ie} & I_{bb} \end{bmatrix} \begin{Bmatrix} F_{ie} \\ F_b \end{Bmatrix} + \begin{bmatrix} \Phi_{in}^T \Gamma_{iN} \\ 0_{bN} \end{bmatrix} \{F_N\} .
$$

(3.30)

Note, the damping coupling partitions are commonly assumed to have the following properties:

$$
[\Phi_{in}^T B_{ii} \Phi_{in}] = [2\xi_n\omega_n], \quad [\Phi_{in}^T B'_{ib}] = [0_{nb}], \quad [B'_{bi} \Phi_{in}] = [0_{bn}].
$$

(3.31)

### 3.2.3  MODAL EFFECTIVE MASS AND TARGET MODE SELECTION

The mass coupling partitions, called modal partition factors are defined as,

$$
[P_{nb}] = [\Phi_{in}^T M'_{ib}], \qquad [P_{bn}] = [P_{nb}]^T .
$$

(3.32)

The modal effective mass associated with one individual mode "$n$" is defined as,

$$
[Meff_{bb}]_n = [P_{bn}] [P_{nb}] .
$$

(3.33)

The product of the above two modal participation matrices is called the modal effective mass for mode "$n$". The sum of all modal effective masses approaches the boundary mass partition, i.e.,

$$
\sum_n [Meff_{bb}]_n \rightarrow [M'_{bb}] .
$$

(3.34)

A linear structure excited only by boundary acceleration has modal accelerations that are stimulated by allocations proportional to the modal participation factors. A convenient form that is commonly employed to evaluate the significance of individual modal responses (i.e., target mode selection) is normalized modal effective mass, specifically defined:

a. total boundary mass is a "row" matrix composed of diagonal terms of $[M'_{bb}]$ (see Equation (3.30));

b. modal effective mass (for mode, "$n$") composed of diagonal terms of $[Meff_{bb}]_n$ (see Equation (3.33)); and

c. normalized modal effective mass (for mode, "$n$") is formed by the quotient $[Meff_{bb}]_n / [M'_{bb}]$.

In the most general case of a structural dynamic model that has a redundant boundary (typically more than 6 DOF), application of the above scheme may be difficult to interpret. However, in most applications, the boundary is constrained to behave as a rigid body referenced to a convenient grid location (6 DOF). In those situations, the total boundary mass conforms to the system's rigid body mass referenced at the selected grid location.

## 3.2.4   ILLUSTRATIVE EXAMPLE: ISS P5 SHORT SPACER

An evaluation of FEM modal content, including modal effective mass, was conducted with results for the lowest 18 FEM modes summarized in Table 3.3.

Table 3.3: ISS P5 short spacer test article FEM modal kinetic energy and modal effective mass

| Mode | Freq (Hz) | Directional KE (%) | | | Directional KE (%) | | | Modal Effective Mass (%) [restricted to Truss & Grapple] | | | | | | Comments |
|---|---|---|---|---|---|---|---|---|---|---|---|---|---|---|
| | | X | Y | X | Truss | Grapple | Fixture | X | Y | Z | ΘX | ΘY | ΘZ | |
| 1 | 18.67 | 94 | 2 | 4 | 56 | 44 | | 49 | | | | 14 | | |
| 2 | 19.31 | 10 | 89 | 1 | 93 | 7 | | | 60 | | 61 | | 64 | |
| 3 | 24.88 | 84 | 4 | 12 | 56 | 44 | | 15 | 1 | 1 | 1 | 1 | 1 | |
| 4 | 28.68 | 46 | 15 | 39 | 75 | 25 | | 1 | 1 | | 2 | | 1 | |
| 5 | 29.37 | 35 | 24 | 41 | 63 | 37 | | 1 | 5 | 6 | 5 | 7 | 4 | |
| 6 | 30.07 | 14 | 23 | 63 | 39 | 61 | | | 5 | 5 | 6 | 3 | 4 | Modes with Modal Effective |
| 7 | 34.01 | 49 | 31 | 20 | 96 | 4 | | | | 1 | | 1 | | Mass >5% are Colored Yellow |
| 8 | 34.65 | 12 | 23 | 65 | 99 | 1 | | | | | | | | |
| 9 | 35.22 | 42 | 51 | 7 | 77 | 23 | | | | | | | | |
| 10 | 35.61 | 40 | 25 | 36 | 89 | 11 | | | | 1 | | | | |
| 11 | 37.82 | 27 | 51 | 22 | 56 | 44 | | | | 2 | | 2 | 2 | |
| 12 | 40.82 | 7 | 18 | 75 | 32 | 68 | | | | 1 | | 1 | | |
| 13 | 42.24 | 6 | 2 | 93 | 16 | 84 | | | | 4 | | 3 | | |
| 14 | 45.21 | 22 | 11 | 67 | 91 | 7 | 2 | 1 | | | 9 | | 6 | |
| 15 | 47.4 | 26 | 52 | 22 | 95 | 4 | 1 | 1 | | | 1 | | 2 | Modes with Non-negligible |
| 16 | 50.22 | 6 | 67 | 27 | 19 | 4 | 76 | | | | 1 | | 1 | Test Fixture KE |
| 17 | 50.41 | 10 | 77 | 13 | 14 | | 86 | | | | | | | Colored Gray |
| 18 | 50.95 | 66 | 23 | 11 | 87 | | 13 | 1 | | 1 | | 1 | | |

Relying on modal effective mass, the target mode selection criterion (modal effective mass > 5%) indicates that only five modes should have been judged "significant." In the conduct of the ISS P5 test program, the first 10 modes were selected as target modes on the basis of NASA's position that the dynamic environment cutoff for the Space Shuttle was $f^* = 35$ Hz.

### 3.2.5 MODAL GAINS AND TARGET MODE SELECTION

By mathematically viewing boundary accelerations, external and nonlinear forces as excitation sources (on a mode-by-mode basis), Equations (3.30) and (3.31) are recast as follows:

$$
\{\ddot{q}_n\} + [2\xi_n\omega_n]\{\dot{q}_n\} + [\omega_n^2]\{q_n\}
= -[P_{nb}]\{\ddot{U}_b\} + [\Phi_{in}^T\Gamma_{ie}]\{F_{ie}\} + [\Phi_{in}^T\Gamma_{iN}]\{F_{iN}\},
\tag{3.35}
$$

$$
\{F_b\} = [M'_{bb}]\{\ddot{U}_b\} + [B'_{bb}]\{\dot{U}_b\} + [K'_{bb}]\{U_b\} + [P_{bn}]\{\ddot{q}_n\} - [\Psi_{ib}^T\Gamma_{ie}]\{F_{ie}\}. \tag{3.36}
$$

The role of modal participation factors, $[P_{nb}]$ and $[P_{bn}]$, and modal effective mass, associated with boundary acceleration excitation was described and demonstrated in the past two sections of this chapter. Equation (3.35) highlights two additional modal metrics, namely the modal gains, $[\Phi_{in}^T\Gamma_{ie}]$ and $[\Phi_{in}^T\Gamma_{iN}]$, which indicate the degree of modal excitation of each particular mode associated with each applied and nonlinear force pattern). The relative magnitudes of these modal gains therefore may serve as metrics for target mode selection in the presence of applied external loads and localized nonlinear forces. When many separate applied force patterns and local nonlinearities are present, modal gain metrics may offer an overly complicated and perplexing target mode selection option.

### 3.2.6 THE MODE ACCELERATION METHOD AND TARGET MODE SELECTION

Consider the following recasting of the upper partition of the matrix structural dynamic Equation (3.28):

$$
[K_{ii}]\{U'_i\} = -[M_{ii}]\{\ddot{U}'_i\} - [B_{ii}]\{\dot{U}'_i\} - [M'_{ib}]\{\ddot{U}_b\} - [B'_{ib}]\{\dot{U}_b\}
- [K'_{ib}]\{U_b\} + [\Gamma_{ie}]\{F_{ie}\} + [\Gamma_{iN}]\{F_{iN}\}. \tag{3.37}
$$

In 1945, Williams [2] introduced the ingenious mode acceleration method, which revolutionized computation of structural dynamic loads process (the benefit of his formulation will be noted later in the present discussion). Since the contribution of damping forces is often small compared to inertial forces, Equation (3.37) becomes

$$
[K_{ii}]\{U'_i\} \approx -[M_{ii}]\{\ddot{U}'_i\} - [M'_{ib}]\{\ddot{U}_b\}
- [K'_{ib}]\{U_b\} + [\Gamma_{ie}]\{F_{ie}\} + [\Gamma_{iN}]\{F_{iN}\}. \tag{3.38}
$$

The final key step in William's development of the mode acceleration method involves substitution of the modal accelerations (see Equation (3.30)) for the physical accelerations, $\{\ddot{U}'_i\}$, resulting in

$$
[K_{ii}]\{U'_i\} \approx -[M_{ii}\Phi_{in}]\{\ddot{q}_n\} - [M'_{ib}]\{\ddot{U}_b\}
- [K'_{ib}]\{U_b\} + [\Gamma_{ie}]\{F_{ie}\} + [\Gamma_{iN}]\{F_{iN}\}. \tag{3.39}
$$

The right-hand side contributions are grouped into "modal acceleration (dynamic)" and "quasi-static" structural forces, which are defined as

$$\{F_i\}_D \approx -[M_{ii}\Phi_{in}]\{\ddot{q}_n\} \quad \text{(modal acceleration forces)}, \tag{3.40}$$

$$\{F_i\}_{QS} \approx -[M'_{ib}]\{\ddot{U}_b\} - [K'_{ib}]\{U_b\} \\ + [\Gamma_{ie}]\{F_{ie}\} + [\Gamma_{iN}]\{F_{iN}\} \quad \text{(quasi-static forces)}. \tag{3.41}$$

The unique contributions associated with William's mode acceleration method are as follows.

1. The quasi-static contribution of all higher modes that are not included in the lower frequency subset defined by $f^*$ are included *without need for their computation!*

2. The coefficient terms associated with modal accelerations, $\{\ddot{q}_n\}$, are associated with *dynamic characteristics of system response*, while the remaining coefficient terms are associated with *quasi-static characteristics of system response*.

Further development of William's mode acceleration method, especially in the aerospace community, requires expansion of the physical solution, Equation (3.39) in terms of absolute displacements employing Equation (3.27), resulting in

$$\begin{Bmatrix} U_i \\ U_b \end{Bmatrix} \approx \begin{bmatrix} -K_{ii}^{-1}M_{ii}\Phi_{in} \\ 0_{bn} \end{bmatrix}\{\ddot{q}_n\} + \begin{bmatrix} -K_{ii}^{-1}M'_{ib} \\ 0_{bn} \end{bmatrix}\{\ddot{U}_b\} \\ + \begin{bmatrix} -K_{ii}^{-1}K'_{ib} \\ I_{bb} \end{bmatrix}\{U_b\} + \begin{bmatrix} K_{ii}^{-1}\Gamma_{ie} \\ I_{be} \end{bmatrix}\{F_{ie}\} + \begin{bmatrix} K_{ii}^{-1}\Gamma_{iN} \\ I_{bN} \end{bmatrix}\{F_{iN}\}. \tag{3.42}$$

The above result implies that the interior stiffness matrix, $[K_{ii}]$, is non-singular. When the stiffness matrix is singular, additional operations are required to achieve the above result.

Stress-strain recovery relationships for detailed structural loads are summarized by

$$\{\sigma\} = [K_\sigma]\{U\}, \quad \{U\} = \begin{Bmatrix} U_i \\ U_b \end{Bmatrix}. \tag{3.43}$$

By combining Equations (3.42) and (3.43), the mode acceleration-based dynamic loads recovery relationship, with load transformation matrices, takes the form

$$\{\sigma\} = [LTM_{\ddot{q}}]\{\ddot{q}_n\} + [LTM_{\ddot{U}_b}]\{\ddot{U}_b\} \\ + [LTM_{U_b}]\{U_b\} + [LTM_{F_e}]\{F_e\} + [LTM_{F_N}]\{F_N\}, \tag{3.44}$$

where the load transformation matrices are

$$[LTM_{\ddot{q}}] = [K_\sigma] \cdot \begin{bmatrix} -K_{ii}^{-1} M_{ii} \Phi_{in} \\ 0_{bn} \end{bmatrix}, \quad [LTM_{\ddot{U}_b}] = [K_\sigma] \cdot \begin{bmatrix} -K_{ii}^{-1} M'_{ib} \\ 0_{bn} \end{bmatrix},$$

$$[LTM_{U_b}] = [K_\sigma] \cdot \begin{bmatrix} -K_{ii}^{-1} K'_{ib} \\ I_{bb} \end{bmatrix}, \tag{3.45}$$

$$[LTM_{F_e}] = [K_\sigma] \cdot \begin{bmatrix} K_{ii}^{-1} \Gamma_{ie} \\ I_{be} \end{bmatrix}, \quad \text{and} \quad [LTM_{F_N}] = [K_\sigma] \cdot \begin{bmatrix} K_{ii}^{-1} \Gamma_{iN} \\ I_{bN} \end{bmatrix}.$$

It should be noted that the above derived form of William's mode acceleration method is one among many other specific forms employed throughout the aerospace community. In particular, forms including damping effects, and more convenient forms associated with substructure (component mode synthesis) and fully assembled vehicle systems are prevalent in the aerospace community. That being said, delineation of accumulated (1) modal acceleration (dynamic) and (2) quasi-static components is most relevant to the topic of target mode selection, specifically, the modal acceleration and quasi-static components, respectively, are

$$\{\sigma\}_D = [LTM_{\ddot{q}}] \{\ddot{q}_n\}, \tag{3.46}$$

$$\{\sigma\}_{QS} = [LTM_{\ddot{U}_b}] \{\ddot{U}_b\} + [LTM_{U_b}] \{U_b\} + [LTM_{F_e}] \{F_e\} + [LTM_{F_N}] \{F_N\}. \tag{3.47}$$

Comprehensive mode acceleration based for target mode selection exploits evaluation of modal and quasi-static contributions to (a) structural forces delineated in Equations (3.40) and (3.41), and/or (b) a representative set of system dynamic loads delineated in Equations (3.46) and (3.47) (termed by some organizations "watch list" loads).

### 3.2.7   ILLUSTRATIVE EXAMPLE: AXISYMMETRIC SHELL FINITE ELEMENT MODEL

The following demonstration analysis is applied to the axisymmetric shell FEM, interpreting the results described in Figure 3.7 and Table 3.2, as associated with a 1/20th sub-scale model. Therefore, the full-scale frequencies of the finite element model are 1/20th of the sub-scale model.

Consider four applied dynamic load distributions, $[\Gamma_{ie}]$, applied in the two lateral directions, the axial direction, and locally normal to the shell wall, respectively (applied boundary accelerations are not included in this exercise), as depicted in Figure 3.9.

In addition, the time history force, $F_e$, along with its normalized shock spectrum (Figure 3.10) is imposed separately on the applied load distributions,

The normalized shock spectrum associated with the applied force time history suggests a full-scale value for $f^*$, the dynamic response cut-off frequency, on the order of 70–80 Hz.

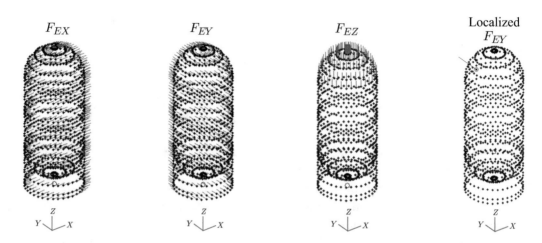

Figure 3.9: Applied load distributions.

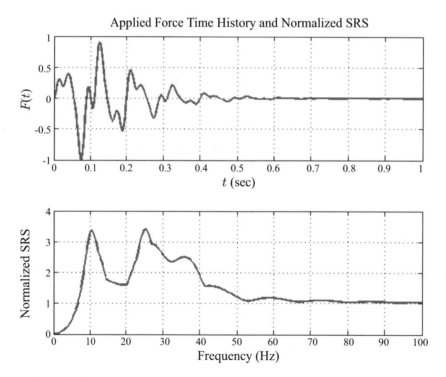

Figure 3.10: Applied force time history and normalized shock spectrum.

The 150th mode of the shell finite element model is about 73 Hz (corresponding to the 1/20th sub-scale value of 1452 Hz).

For the present example, the modal response equation (see Equation (3.35)) reduces to,

$$\{\ddot{q}_n\} + [2\xi_n\omega_n]\{\dot{q}_n\} + [\omega_n^2]\{q_n\} = [\Phi_n^T\Gamma_e]\{F_e\}. \tag{3.48}$$

Therefore the modal gains for each of the four applied load distributions are $[\Phi_n^T\Gamma_e]$.

Regarding the first three applied load distributions described in Figure 3.5, the dynamic loads, $\{\sigma\}$, of interest are the cumulative "body" loads at a series of 26 axial stations, defined by the following particular mode acceleration equations:

$$\{\sigma\} = \{\sigma\}_D + \{\sigma\}_{QS} = [LTM_{\ddot{q}}]\{\ddot{q}_n\} + [LTM_{F_e}]\{F_e\},$$
$$[LTM_{\ddot{q}}] = [\Psi_b^T][M][\Phi], \quad [LTM_{F_e}] = [\Psi_b^T][\Gamma_e]. \tag{3.49}$$

For the fourth applied load distribution case involving a single concentrated load and an associated coinciding local response load of interest is described by the load transformation matrices,

$$[LTM_{\ddot{q}}] = m_{DOF}[\Phi_{DOF}], \quad [LTM_{F_e}] = 1, \tag{3.50}$$

where the "DOF" subscript denotes the local matrix components of interest.

An evaluation of mode-by-mode contributions to dynamic response for all four applied load distribution cases is summarized in Figure 3.11. (It should be noted that the results are associated with "unit" values for all modal accelerations.) For each of the four applied load distributions, there are groupings of three plots; the top plot indicates the distribution of modal frequencies, the middle plot summarizes relative modal gain norms, and the lower plot indicates an absolute summation "norm" of all dynamic loads relevant for the particular applied load distribution case.

The following conclusions are drawn from the above results.

1. Inertial "body" load responses, representative of "primary" structure loads, associated with the first three applied load distributions, are the result of contributions associated with "body" modes (24 out of a total of 150 modes).

2. Local inertial load response, representative of "secondary" structure loads, associated with the fourth applied load case is the result of many of the 150 "body" and "breathing" modes.

3. In all cases, modal gains offer an indication of the relative prominence of individual mode contributions, but the dynamic response load norms for $\{\sigma\}_D$ offer a more definitive assessment of the significant modal contributions to a responding loads-modes "watch list."

4. It appears that predicted system dynamic response "decompositions" are the most relevant discriminators for target mode selection. Modal gain and modal effective mass may also serve as less definitive target mode selection indicators.

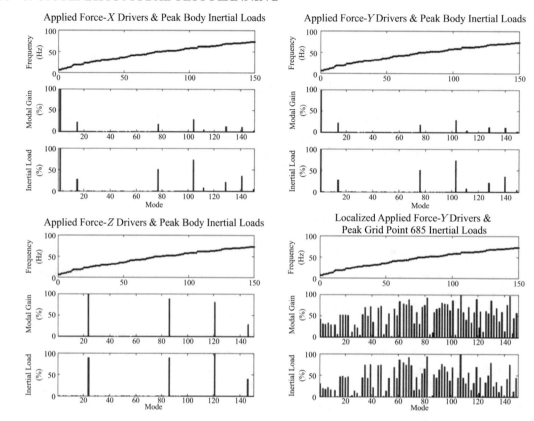

Figure 3.11: Mode-by-mode contributions to dynamic response.

The relative significance of quasi-static and modal dynamic contributions to system transient responses, for the first three applied load distribution cases, is summarized in Table 3.4.

In all three distributed applied load distribution cases, the quasi-static contributions to system response are generally significant. However, the fourth concentrated applied load response case, depicted in Figure 3.12, is most interesting in that the local responding load is almost completely dominated by quasi-static response (in spite of the fact that many modes contribute to the minor, dynamic component of response.

While the results for the fourth applied load case are anticlimactic, (i.e., (a) dynamic response is associated with the contributions of many modes and (b) total response is nearly entirely due to quasi-static response), the vital role played by Williams' mode acceleration method is most clearly evident. Specifically:

1. system loads due to applied dynamic environments are the sum of "quasi-static" and "modal dynamic" contributions;

Table 3.4: Significance of quasi-static and modal dynamic contributions to system response

| Load Case 1: Distributed "X" Loading | | | | | | Load Case 2: Distributed "Y" Loading | | | | | | Load Case 3: Distributed "Z" Loading | | | | | |
|---|---|---|---|---|---|---|---|---|---|---|---|---|---|---|---|---|---|
| Geometry | | \|FX\| | | \|MY\| | | Geometry | | \|FY\| | | \|MX\| | | Geometry | | \|FZ\| | | \|FR\| | |
| R | Z | Static | Total | Static | Total | R | Z | Static | Total | Static | Total | R | Z | Static | Total | Static | Total |
| 400 | 0 | 15 | 10 | | 47 | 400 | 0 | 15 | 10 | | 47 | 400 | 0 | | 47 | | 4 |
| 400 | -60 | 15 | 12 | | 37 | 400 | -60 | 15 | 12 | | 37 | 400 | -60 | | 32 | | 7 |
| 400 | -200 | 15 | 13 | | 40 | 400 | -200 | 15 | 13 | | 40 | 400 | -200 | | 33 | | 12 |
| 400 | -340 | 15 | 14 | | 25 | 400 | -340 | 15 | 14 | | 25 | 400 | -340 | | 15 | | 11 |
| 400 | -480 | 15 | 15 | | 6 | 400 | -480 | 15 | 15 | | 7 | 400 | -480 | | 3 | | 5 |
| 400 | -540 | 15 | 15 | | 13 | 400 | -540 | 15 | 15 | | 13 | 400 | -540 | | 1 | | |
| 392 | -80 | | 9 | | 37 | 392 | -80 | | 9 | | 37 | 392 | -80 | | 21 | | 4 |
| 352 | -190 | | 10 | | 36 | 352 | -190 | | 10 | | 36 | 352 | -190 | | 39 | | 3 |
| 264 | -300 | | 8 | | 16 | 264 | -300 | | 8 | | 16 | 264 | -300 | | 35 | | 1 |
| 152 | -370 | | 4 | | 3 | 152 | -370 | | 4 | | 3 | 152 | -370 | | 19 | | |
| 40 | -398 | | 1 | | 0 | 40 | -398 | | 1 | | | 40 | -398 | | 4 | | |
| 400 | 60 | 15 | 9 | | 44 | 400 | 60 | 15 | 9 | | 44 | 400 | 60 | | 35 | | 6 |
| 400 | 200 | 15 | 8 | | 72 | 400 | 200 | 15 | 8 | | 72 | 400 | 200 | | 61 | | 9 |
| 400 | 340 | 15 | 10 | | 81 | 400 | 340 | 15 | 10 | | 81 | 400 | 340 | | 71 | | 8 |
| 400 | 480 | 15 | 5 | | 62 | 400 | 480 | 15 | 5 | | 62 | 400 | 480 | | 56 | | 5 |
| 400 | 540 | 15 | 6 | | 43 | 400 | 540 | 15 | 6 | | 43 | 400 | 540 | | 34 | | 3 |
| 400 | 600 | 15 | 9 | | 64 | 400 | 600 | 15 | 9 | | 64 | 400 | 600 | | 61 | | 4 |
| 400 | 740 | 15 | 22 | | 96 | 400 | 740 | 15 | 22 | | 97 | 400 | 740 | | 90 | | 6 |
| 400 | 880 | 15 | 26 | | 100 | 400 | 880 | 15 | 26 | | 100 | 400 | 880 | | 95 | | 5 |
| 400 | 1020 | 15 | 17 | | 75 | 400 | 1020 | 15 | 17 | | 75 | 400 | 1020 | | 71 | | 5 |
| 400 | 1080 | 15 | 11 | | 52 | 400 | 1080 | 15 | 11 | | 52 | 400 | 1080 | | 47 | | 5 |
| 392 | 1160 | 15 | 18 | | 67 | 392 | 1160 | 15 | 18 | | 67 | 392 | 1160 | 62 | 81 | | 7 |
| 352 | 1270 | 15 | 25 | | 65 | 352 | 1270 | 15 | 25 | | 65 | 352 | 1270 | 62 | 100 | | 6 |
| 264 | 1380 | 15 | 20 | | 30 | 264 | 1380 | 15 | 20 | | 30 | 264 | 1380 | 62 | 90 | | 1 |
| 152 | 1450 | 15 | 11 | | 6 | 152 | 1450 | 15 | 11 | | | 152 | 1450 | 62 | 72 | | |
| 40 | 1478 | 15 | 12 | | 0 | 40 | 1478 | 15 | 12 | | | 40 | 1478 | 62 | 64 | | |

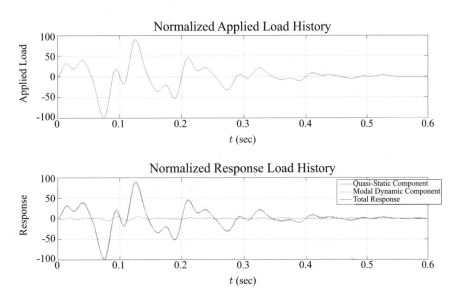

Figure 3.12: Quasi-static and modal dynamic contributions to system response (load case 4).

2. there are situations in which response loads may be almost entirely the result of "quasi-static" behavior; hence precise knowledge of modal dynamics is unnecessary; and

3. in the majority of situations for which modal dynamics play a significant role, response loads (especially primary structural loads) are the product of some system mode contri-

butions; in these situations, plans for modal testing are best focused on a selected "target mode" subset.

### 3.2.8   CLOSURE

A variety of target mode selection metrics for modal test planning have been explored. When a subject system is expected to experience dynamic environments imposed as interface dynamic motions (no other applied load sources), modal effective mass appears to provide an adequate metric for target mode selection, as employed in the ISS P5 modal test plan and specified in NASA STD-5002. For the more general case of a subject system expected to experience distributed applied forces (and in some cases interface dynamic motions), a detailed structural stress and member response load (also called a loads "watch list") modal decomposition provides the most reliable metric for target mode selection. The "watch list" modal decomposition process relies on Williams' mode acceleration method and load decomposition matrices that categorize stress and member response load contributions in (a) quasi-static and (b) modal dynamic categories. As demonstrated with the axisymmetric shell finite element model, the "watch list" decomposition process offers an effective strategy for dealing with the challenging "many modes" problem.

## 3.3   PART 3: RESPONSE DOF SELECTION FOR MAPPING EXPERIMENTAL MODES

### 3.3.1   INTRODUCTION

The Residual Kinetic Energy (RKE) method for response "measurement" DOF selection to map experimental normal modes is described and demonstrated on a simple rod finite element modal and a pre-test FEM associated with a modal test article. The method utilizes a detailed pre-test finite element model of the subject structure. An initial response "measurement" DOF set is first selected for mapping modes in the frequency band of interest. The remaining DOFs of the structural model are assumed to respond in a quasi-static manner following a Guyan reduction transformation. Modes formed from the "measurement" partition of the FEM modal vectors and expanded using the Guyan reduction transformation comprise the "approximate" normal mode set. While orthogonality of the complete FEM is mathematically perfect, orthogonality associated with the approximate normal mode set is necessarily imperfect. The generally accepted standard for U.S. government sponsored aerospace systems is specified by NASA and the USAF Space Command. Residual error vectors are formed based on differences between complete FEM normal modes in the frequency band of interest and the set of corresponding "approximate" modes. A RKE matrix directly identifies additional DOFs (or DOF groupings) which must be instrumented in order to experimentally map the normal modes of interest.

## 3.3.2    SELECTION OF DOFS FOR MODAL MAPPING (THE RKE METHOD)

The most commonly used approach for development of a TAM is Guyan eduction [3] in which the "measurement" DOFs are retained in the "analysis" set. The matrix dynamic equations for free vibration (with all physical constraints applied) are

$$[K]\{\Phi\} - [M]\{\Phi\}\lambda = \{0\} \rightarrow \begin{bmatrix} K_{AA_1} & K_{AO} \\ K_{OA} & K_{OO} \end{bmatrix} \begin{Bmatrix} \Phi_A \\ \Phi_O \end{Bmatrix}$$
$$- \begin{bmatrix} M_{AA_1} & M_{AO} \\ M_{OA} & M_{OO} \end{bmatrix} \begin{Bmatrix} \Phi_A \\ \Phi_O \end{Bmatrix} \lambda = \begin{Bmatrix} 0 \\ 0 \end{Bmatrix}, \tag{3.51}$$

where the FEM modal matrix is expressed in terms of "analysis" and "omitted" partitions. Denoting the modal matrix as, $[\Phi]$, perfect orthogonality of (unit mass normalized) modes is expressed as

$$[OR] = [\Phi]^T [M] [\Phi] = [I]. \tag{3.52}$$

The Guyan reduction transformation which is used to form an approximate test-analysis mass (TAM) mass matrix is

$$[T_{RED}] = \begin{bmatrix} I \\ -K_{OO}^{-1} K_{OA} \end{bmatrix}. \tag{3.53}$$

If the "analysis" set partition of the modal matrix is assumed to represent "measured" modal DOFs, then the approximate TAM modal matrix is

$$\begin{bmatrix} \Phi_A \\ \Psi_O \end{bmatrix} = \begin{bmatrix} I \\ -K_{OO}^{-1} K_{OA} \end{bmatrix} [\Phi_A], \tag{3.54}$$

where the partition, $\Psi_O$, represents the approximate "omitted" DOF partition. Moreover, the TAM mass matrix, $[M_{AA}]$, is defined as

$$[M_{AA}] = \begin{bmatrix} I \\ -K_{OO}^{-1} K_{OA} \end{bmatrix}^T \begin{bmatrix} M_{AA_1} & M_{AO} \\ M_{OA} & M_{OO} \end{bmatrix} \begin{bmatrix} I \\ -K_{OO}^{-1} K_{OA} \end{bmatrix}. \tag{3.55}$$

Imperfect orthogonality of the modal partition, $[\Phi_A]$, corresponding to the "measured" DOFs with respect to the TAM mass matrix (for unit mass normalized modes, $[\Phi_A]$) is expressed as,

$$[OR_A] = [\Phi_A]^T [M_{AA}] [\Phi_A] \neq [I_{AA}]. \tag{3.56}$$

The U.S. government test mode orthogonality [1, 4] criterion requires $|OR_{A,ij}| \leq 10\%$ for all $i \neq j$ terms in the TAM orthogonality matrix, $[OR_A]$.

The residual displacement error matrix based on the difference between the exact FEM and TAM (Equation (3.54)) modal matrices is defined as

$$[R] = \begin{bmatrix} \Phi_A \\ \Phi_O \end{bmatrix} - \begin{bmatrix} \Phi_A \\ \Psi_O \end{bmatrix} = \begin{bmatrix} 0 \\ \Phi_O - \Psi_O \end{bmatrix}. \tag{3.57}$$

Note that the residual error associated with the "measured" or "analysis" DOF partition is null. The modal kinetic energy distribution for the complete system is

$$[E_\Phi] = [M\Phi] \otimes [\Phi], \tag{3.58}$$

where the column sum for each individual mode is unity (if the modes are normalized to unit modal mass).

The RKE matrix is defined in a similar manner as

$$[E_R] = [MR] \otimes [R]. \tag{3.59}$$

Like the residual displacement error, $[R]$, the RKE matrix is exactly $[0]$ at the rows corresponding to measured DOFs. The expected characteristic that residual energy is pronounced at "omitted" yet dynamically significant DOFs in any particular mode will be demonstrated with the uniform, free-free rod described by a 19-grid point FEM, illustrated in Figure 3.13.

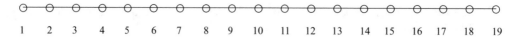

Figure 3.13: Simple rod structure.

The first six (6) axial vibration modes are graphically depicted in Figure 3.14.

DOFs 1, 3, 5, 7, 9, and 19 were first selected as "measured" DOFs. The TAM orthogonality matrix and RKE associated with this selected DOF set are illustrated in Figure 3.15.

The orthogonality matrix and RKE plot indicate that the selected "measured" DOFs are inadequate for mapping of modes 3–6. Pronounced RKE in DOFs 11–17 suggests addition of "measured" DOFs in modes that vicinity.

Addition of accelerometers at DOFs 11, 13, 15, and 17 was found to improve both the orthogonality matrix (conforming to the NASA-STD-5002 and SMC-S-004 criteria) and RKE, as indicated in Figure 3.16.

Since its introduction in 1998 [5], the RKE method has been employed by many organizations in the U.S. aerospace industry for modal test planning. A variety of enhancements by a number of authors, involving the application of iterative and genetic schemes and modifications to the basic Guyan reduction method, have been introduced over the years. However all of the enhancements are based on the original RKE formulation.

### 3.3.3   ILLUSTRATIVE EXAMPLE: ISS P5 SHORT SPACER

The RKE method was employed for determination of accelerometer allocation for the ISS P5 Short Spacer modal test program. An allocation of 86 triaxial accelerometers (258 DOF) was selected, resulting in a TAM mass matrix that produced impressive orthogonality for the majority of 36 FEM modes, as indicated in Figure 3.17.

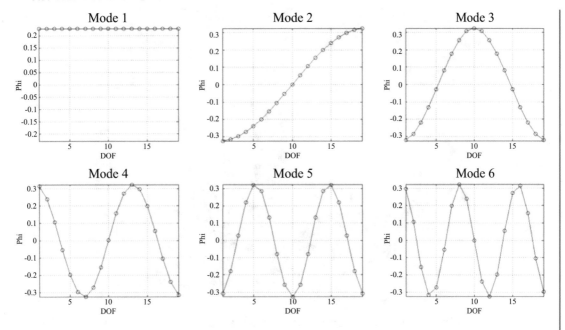

Figure 3.14: **Simple bar structure modes.**

| | |Orthogonality| (%) | | | | |
|---|---|---|---|---|---|---|
| Mode | 1 | 2 | 3 | 4 | 5 | 6 |
| 1 | 100 | 10 | 14 | 67 | 57 | 6 |
| 2 | 10 | 100 | 6 | 45 | 48 | 35 |
| 3 | 14 | 6 | 100 | 20 | 19 | 58 |
| 4 | 67 | 45 | 20 | 100 | 40 | 26 |
| 5 | 57 | 48 | 19 | 40 | 100 | 12 |
| 6 | 6 | 35 | 58 | 26 | 12 | 100 |

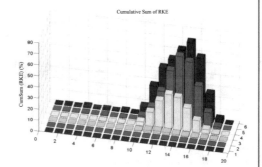

Figure 3.15: **TAM orthogonality and RKE for the 6 DOF accelerometer set.**

| | |Orthogonality| (%) | | | | |
|---|---|---|---|---|---|---|
| Mode | 1 | 2 | 3 | 4 | 5 | 6 |
| 1 | 100 | 0 | 0 | 0 | 1 | 0 |
| 2 | 0 | 100 | 0 | 1 | 0 | 2 |
| 3 | 0 | 0 | 100 | 0 | 2 | 0 |
| 4 | 0 | 1 | 0 | 100 | 0 | 3 |
| 5 | 1 | 0 | 2 | 0 | 100 | 0 |
| 6 | 0 | 2 | 0 | 3 | 0 | 100 |

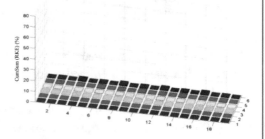

Figure 3.16: **TAM orthogonality and RKE for the 10 DOF accelerometer set.**

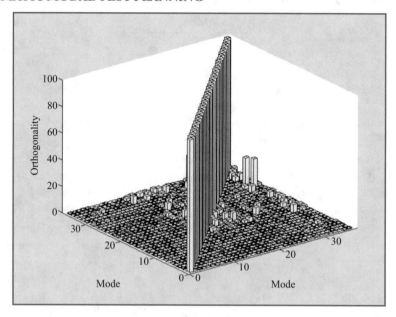

Figure 3.17: ISS P5 pre-test modal orthogonality prediction.

Table 3.5: Summary of pre-test modal characteristics and TAM mass matrix modal orthogonality

| Mode | Freq (Hz) | Kinetic Energy (%) | | | \|Orthogonality\| (%) | | | | | | | | | | | | | | | | |
|---|---|---|---|---|---|---|---|---|---|---|---|---|---|---|---|---|---|---|---|---|---|
| | | Truss | Grapple | Fixture | | | | | | | | | | | | | | | | | |
| 1 | 18.67 | 56 | 44 | | 100 | | | | | | 1 | | | | | | | | | | |
| 2 | 19.31 | 93 | 7 | | | 100 | | | | | | | | | | | | | | | |
| 3 | 24.88 | 56 | 44 | | | | 100 | | | | | | | | | | | | | | |
| 4 | 28.68 | 75 | 25 | | | | | 100 | | | | | | | | | | 1 | | | |
| 5 | 29.37 | 63 | 37 | | | | | | 100 | | | | | | | | | | | | |
| 6 | 30.07 | 39 | 61 | | | | | | | 100 | | | | 1 | | | | | 1 | | |
| 7 | 34.01 | 96 | 4 | | | | | | | | 100 | | 1 | | | | | | | | |
| 8 | 34.65 | 99 | 1 | | | | | | | | | 100 | | | | | | | | | |
| 9 | 35.22 | 77 | 23 | | | | | | | 1 | | | 100 | 1 | 1 | | | | | | |
| 10 | 35.61 | 89 | 11 | | | | | | | | | | 1 | 100 | | | | | | | |
| 11 | 37.82 | 56 | 44 | | | | | | | | | | 1 | | 100 | | 1 | | | | |
| 12 | 40.82 | 32 | 68 | | | | | | 1 | | | | | 100 | 1 | 100 | 1 | | | 1 | |
| 13 | 42.24 | 16 | 84 | | | | | | | | | | | 1 | 1 | 100 | 1 | | 1 | | |
| 14 | 45.21 | 91 | 7 | 2 | | | | 1 | | | | | | 1 | 100 | | | 3 | 1 |
| 15 | 47.4 | 95 | 4 | 1 | | | | | | | | | | | | | 100 | 3 | 1 |
| 16 | 50.22 | 19 | 4 | 76 | | | | | | 1 | | | | 1 | 1 | 3 | 3 | 100 | 1 |
| 17 | 50.41 | 14 | | 86 | | | | | | | | | | | 1 | 1 | 1 | 1 | 100 |

A more detailed summary justifying the RKE selected TAM mass matrix and accelerometer allocation plan (for predicted modes 1–17) is provided in Table 3.5.

The above summary suggests that the RKE-derived modal test plan includes sufficient accelerometer channels to map modes (associated with the P5 test article and test fixture) in the 0–50 Hz frequency band, while satisfying NASA-STD-5002 TAM orthogonality criteria. Note modes 14–17 include non-negligible modal kinetic energy content in the test fixture.

### 3.3.4    ALLOCATION OF MODAL EXCITATION RESOURCES

Planning of excitation resources for a modal test involves placement and orientation of shakers based on

1. accessibility and structural integrity of candidate test article locations; and

2. ability to adequately excite all target modes (using modal gains, see Equation (3.48)).

Employing the above criteria, diagonally oriented shakers (and in-line accelerometer channels intended for measurement of drive point frequency response functions, to be discussed in Chapter 4) were allocated as depicted in Figure 3.18.

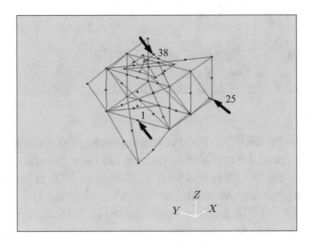

| TAM Node | Description | Location (Coordinates) | | | Direction Cosines (Radians) | | |
|---|---|---|---|---|---|---|---|
| | | $X$ (in) | $Y$ (in) | $Z$ (in) | $Fx/Ftot$ | $Fy/Ftot$ | $Fz/Ftot$ |
| 1 | Lower-Center-Front | 776.38 | 0 | 380.31 | 0.58 | 0.58 | 0.58 |
| 25 | Lower-Left-Rear | 890.33 | -54.5 | 380.31 | 0.58 | 0.58 | 0.58 |
| 38 | Grapple Fixture | 820.24 | 0 | 466.16 | 0.58 | 0.58 | -0.58 |

Figure 3.18: ISS P5 excitation allocation.

The anticipated adequacy of excitation resources was evaluated on the basis of predicted modal gains associated with the test article finite element model, as summarized in Table 3.6. Note that the most prominent shaker location for each individual mode is designated in **bold** print and modes with non-negligible test fixture kinetic energy are shaded.

Table 3.6: Predicted adequacy of ISS P5 modal excitation resources

| Mode | Freq (Hz) | Kinetic Energy (%) | | | Normalized Modal Gain, [$\Phi^T\Gamma$] | | |
|---|---|---|---|---|---|---|---|
| | | Truss | Grapple | Fixture | 1 (Truss) | 25 (Truss) | 38 (Grapple) |
| 1 | 18.67 | 56 | 44 | | 22 | -13 | **77** |
| 2 | 19.31 | 93 | 7 | | -15 | **-26** | -21 |
| 3 | 24.88 | 56 | 44 | | -27 | -6 | **100** |
| 4 | 28.68 | 75 | 25 | | 20 | 52 | **61** |
| 5 | 29.37 | 63 | 37 | | -5 | **10** | -9 |
| 6 | 30.07 | 39 | 61 | | -9 | -21 | **29** |
| 7 | 34.01 | 96 | 4 | | 5 | **45** | 2 |
| 8 | 34.65 | 99 | 1 | | -1 | **16** | -4 |
| 9 | 35.22 | 77 | 23 | | -9 | **18** | -13 |
| 10 | 35.61 | 89 | 11 | | -10 | **-99** | 18 |
| 11 | 37.82 | 56 | 44 | | 12 | -12 | **-67** |
| 12 | 40.82 | 32 | 68 | | 10 | -20 | **51** |
| 13 | 42.24 | 16 | 84 | | **-10** | -7 | 2 |
| 14 | 45.21 | 91 | 7 | 2 | -11 | **-24** | 10 |
| 15 | 47.4 | 95 | 4 | 1 | -54 | **82** | 17 |
| 16 | 50.22 | 19 | 4 | 76 | **-32** | -9 | -26 |
| 17 | 50.41 | 14 | | 86 | **12** | 1 | -2 |

## 3.3.5   CLOSURE

The theoretical basis of the RKE method for accelerometer allocation and TAM mass matrix definition as part of the modal test planning process has been described. An early application of the RKE method for the ISS P5 modal test, conducted in 2001 is summarized in the above discussion. One of many subsequent modal tests conducted in the U.S. aerospace community, specifically NASA MSFC's ISPE modal test, conducted in 2016, is employed as an illustrative example throughout this book. In addition to accelerometer allocation, the allocation of shaker resources based on pre-test FEM predicted modal gains (a common practice) completes the process of accelerometer and excitation allocation specification.

## 3.4   REFERENCES

[1] Load analysis of spacecraft and payloads, *NASA-STD-5002*, 1996. 47, 59

[2] D. Williams, Dynamic loads on aeroplanes under given impulsive load with particular reference to landing and gust loads on a large flying boat, *Great Britain Royal Aircraft Establishment Reports SME 3309 and 3316*, 1945. 51

[3] R. Guyan, Reduction of stiffness and mass matrices, *AIAA Journal*, 3, 1965. DOI: 10.2514/3.2874. 59

[4] Independent structural loads analysis, *U.S. Air Force Space Command, SMC-S-004*, 2008. 59

[5] R. N. Coppolino, Automated response DOF selection for mapping of experimental normal modes, *IMAC XVI*, 1998. 60

# CHAPTER 4

# Measured Data Analysis

## 4.1 PART 1: PRELIMINARY MEASURED DATA ANALYSIS

### 4.1.1 INTRODUCTION

Physical phenomena are either deterministic or random (non-deterministic) [1]. A determin-istic time history record follows an explicit mathematical relationship (e.g., a sinusoid) while a random time history can only be known in terms of statistical characteristics (e.g., mean value, standard deviation). Deterministic time history data classifications are summarized in Table 4.1.

Table 4.1: Deterministic time history data classifications

| Deterministic Time History Records | | |
|---|---|---|
| Class | Sub-Class | Example |
| Periodic | Sinusoidal | $A \cdot \sin(\omega t)$ |
| | Complex Periodic | $\sum\limits_{n=1}^{N} A_n \sin(n\omega_0 t + \lambda_n)$ |
| Non-Periodic | Almost Periodic | $\sum\limits_{n=1}^{N} A_n \sin(\omega_n t + \lambda_n)$ $\omega_{n+1}/\omega_n$ is not an integer |
| | Transient | $\sum\limits_{n=1}^{N} A_n e^{-\sigma_n t} \sin(\omega_n t + \lambda_n)$ |

Several descriptive terms for random data records are now introduced. A collection of time history records from separate "tests" is called an *ensemble*. A single test record is called *stationary* if the statistical properties (e.g., mean, standard deviation, etc.) are the same for all sub-records of "reasonable" duration (selected segments from the entire test record). Otherwise, the record is classified as *non-stationary*. If each data record within an ensemble is stationary and the statistical properties of all data records are the same, the ensemble is classified as *ergodic*. Note that *all ergodic processes are stationary*, but all stationary processes are not ergodic. Random data classifications [1] are summarized in Table 4.2.

Preliminary measured data analysis (which employs basic data analysis functions) is re-quired for two primary reasons, specifically:

Table 4.2: Random time history data classifications

| Random Time History Records | | |
|---|---|---|
| Class | Sub-Class | Example |
| Stationary | Ergodic | All tests have the same standard deviation |
| | Non-Ergodic | All tests are stationary, but individual record "intensities" differ |
| Non-Stationary | See Bendat and Piersol [1] | See Bendat and Piersol [1] |

1. overall measured data quality and content evaluation and

2. detection of nonlinear dynamics for a subject test article.

## 4.1.2   KEY PRELIMINARY MEASURED DATA ANALYSIS FUNCTIONS, MEAN, VARIANCE, AND STANDARD DEVIATION

The mean value, $\mu$, of a time history data record, $[x]$, with "$N$" samples, is calculated within as [1]

$$\mu = (1/N) \sum_{n=1}^{N} x(n),$$  (4.1)

and the (sample) variance, $\sigma^2$, of a time history data record is calculated as [1]

$$\sigma^2 = (1/N) \sum_{n=1}^{N} [x(n) - \mu]^2.$$  (4.2)

The (sample) standard deviation, $\sigma$, is simply the square root of the (sample) variance.

## 4.1.3   NORMALIZED PROBABILITY DENSITY AND IDEAL GAUSSIAN DISTRIBUTION

The normalized probability density function for a measured time history record is calculated based on the normalized data record, $z$, defined as [1]

$$z = (x - \mu)/\sigma.$$  (4.3)

Dividing the normalized data record (composed of "$N$" samples) into "$n$" equally spaced containers or "bins" between the minimum and maximum values of $z$, the histogram function, $h(z)$, is formed as the number of samples within each "bin" divided by "$N$". For a data record, the normalized probability density function is defined as [1];

$$p(z) = \left(\frac{n}{z_{max} - z_{min}}\right) \cdot \left(\frac{h(z)}{N}\right). \tag{4.4}$$

The ideal normalized Gaussian probability density function, $p_G(z)$, is defined as [1];

$$p_G(z) = \left(1/\sqrt{2\pi}\right) \cdot e^{-(z^2/2)}. \tag{4.5}$$

Both the actual normalized probability density and normalized Gaussian probability density functions have the integral property [1]

$$\int_{-\infty}^{\infty} p(z)dz = \int_{-\infty}^{\infty} p_G(z)dz = 1. \tag{4.6}$$

## 4.1.4   TOTAL NORMALIZED PROBABILITY FUNCTION

The total probability that the normalized time history will fall within a specified range is defined as [1],

$$P(z_1 \leq z \leq z_2) = \int_{z_1}^{z_2} p(z)dz, \tag{4.7}$$

and the probability that the value of $z$ will not exceed an upper bound, $z_2$, is defined as

$$P(-\infty \leq z \leq z_2) = \int_{-\infty}^{z_2} p(z)dz. \tag{4.8}$$

## 4.1.5   AUTOSPECTRUM OR POWER SPECTRAL DENSITY FUNCTION

The autospectrum or power spectral density function is defined in the literature by a variety of relationships. For a thorough exposition of this function [1], a physically descriptive definition of the autospectrum is based on the way this quantity was estimated prior to the popular use of digital signal processing techniques. The analog process for estimating autospectrum involved four steps.

1. Filter a time history, $x(t)$, with a narrow bandpass filter of bandwidth, $\Delta f$, and center frequency, $f$ (Hz), resulting in $x(f, \Delta f, t)$.

2. Form the square of the filtered signal, $x^2(f, \Delta f, t)$.

3. Calculate the average value of the squared, filtered signal,

$$\bar{x}^2(f, \Delta f, t) = \frac{1}{T} \int_0^T x^2(f, \Delta f, t)dt. \tag{4.9}$$

4. Divide the result by the filter bandwidth, $\Delta f$, resulting in the autospectrum (or power spectrum),

$$G_{xx}(f) = \frac{\bar{x}^2(f, \Delta f, t)}{\Delta f} = \frac{1}{(\Delta f \cdot T)} \int_0^T x^2(f, \Delta f, t) dt. \qquad (4.10)$$

The (real-valued) autospectrum provides information on the frequency distribution of the standard deviation of a time history record. It therefore has the following property:

$$\sigma^2 = \int_0^\infty G_{xx}(f) df. \qquad (4.11)$$

A modern digital, state-of-the-art method for calculating the autospectrum employs the finite Fourier transform of the time history record. By subdividing the record into "$n_d$" distinct sub-records or "windows" (of duration, $\Delta T$) the autospectrum is defined as,

$$G_{xx}(f) = \frac{2}{(n_d \Delta T)} \cdot \sum_{i=1}^{n_d} X_i(f) * X_i^*(f). \qquad (4.12)$$

$X(f)$ is the finite Fourier transform of the time domain series, $x(t)$, where the temporal sampling rate is $\Delta t$. The autospectrum, $G_{XX}(f)$, is calculated at discrete frequencies, $f_k = k \cdot \Delta f$, where $0 \le k \le n_w/2$. Important guidelines for computation of an appropriate autospectrum are:

$$f_{\max} = \frac{1}{2 \cdot \Delta t} \text{ (Nyquist frequency)}, \quad \Delta f = \frac{1}{\Delta T} \text{ (bandwidth resolution)}$$

$$n_w = \frac{\Delta T}{\Delta t} \text{ (discrete Fourier transforms "window" length)} \qquad (4.13)$$

$$T_{\max} = n_d \cdot \Delta T \text{ (total record length for "}n_d\text{" distinct averages)}.$$

In practice, the originally measured "analog" signal, $x(t)$, should be low-pass filtered with a "cut-off" frequency (typically $0.8 \times f_{\max}$) in order to avoid aliasing before conversion to a time series with digital bandwidth, $f_{\max}$. In addition, a sufficiently long data record, $T_{\max}$, should be recorded to estimate an autospectrum associated with a desired bandwidth resolution, $\Delta f$, and number of distinct averages, $n_d$ (typically in excess of 10 to minimize contributions associated with extraneous noise sources). It should finally be noted that specialized "windowing" functions and correction factors (e.g., the Hanning window and "bow-tie" correction factors) and overlap processing (non-distinct successive records) are commonly applied practices [1].

### 4.1.6   CROSS-SPECTRUM OR CROSS-POWER SPECTRAL DENSITY FUNCTION

Closely related to the autospectrum is the cross-spectrum, which will be featured in Part 2 of this chapter. In a manner similar to Equation (4.13), the digital cross-spectrum is defined as [1]

$$G_{yx}(f) = \frac{2}{(n_d \Delta T)} \cdot \sum_{i=1}^{n_d} Y_i(f) * X_i^*(f). \qquad (4.14)$$

Guidelines and practices for estimation of a digital cross-spectrum and autospectrum are identical.

### 4.1.7   THE SPECTROGRAM FUNCTION

An extremely informative application of the autospectrum, namely the spectrogram, is defined as a "running" autospectrum, which is a function of individual or "grouped" autospectra that are computed at successive time segments of a data record. The value of spectrograms and all of the above described preliminary measured data analysis functions is demonstrated in the following discussion.

A series of illustrative examples are provided to demonstrate application of preliminary data analysis fundamentals in various situations. Overall classification of individual time history records is effected through calculation and display of the time history, spectrogram, autospectrum and probability density function. Data quality and content is evaluated by closer examination of the autospectrum (including the spectrogram), probability density and total probability, and response spectrum.

### 4.1.8   ILLUSTRATIVE EXAMPLE: SINUSOIDAL TIME HISTORY WITH BACKGROUND RANDOM NOISE

Consider a measured time history record consisting of a 20 Hz sinusoidal signal, contaminated by broadband random noise, and sampled at $dt = .005$ s over a duration of 50 s. The composite preliminary data analysis display shown in Figure 4.1 includes a color spectrogram (upper left), time history trace (lower left), autospectrum (upper right), and probability density (lower right). The dashed curve in the probability density plot indicates the ideal Gaussian probability density function. In addition, the title indicates the window length ($N_W$), frequency bandwidth ($\Delta f$), and number of distinct averages ($N_{av}$) of autospectrum sub-records, as well as the mean and standard deviation (Std) values for the data record.

Based on general uniformity of the spectrogram and time history plots with respect to time, the data record is judged *stationary*. Both the spectrogram and autospectrum indicate that the record is dominated by a 20 Hz sinusoidal signal. The record, in spite of the presence of broad band noise, may be classified as predominantly deterministic due to general consistency of the sinusoidal component. The probability density function differs substantially from the ideal Gaussian distribution in a manner suggesting a sinusoid (see [2]).

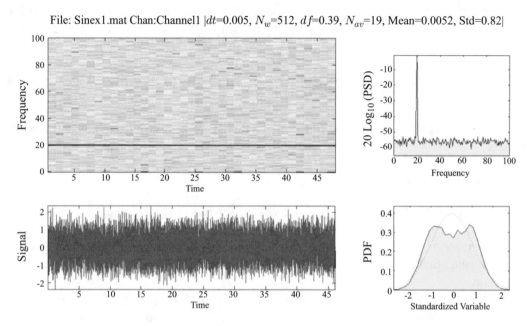

Figure 4.1: Preliminary data analysis of a sinusoidal time history with random noise record.

### 4.1.9    ILLUSTRATIVE EXAMPLE: SDOF LINEAR SYSTEM RESPONSE TO RANDOM EXCITATION

An SDOF linear system, described by the second-order dynamic equation,

$$\ddot{u}(t) + 2\zeta_n \omega_n \dot{u}(t) + \omega_n^2 u(t) = F(t), \tag{4.15}$$

is subjected to broad band Gaussian random excitation, $F(t)$, and has the displacement response, $u(t)$. Both excitation and response are sampled at $dt = .005$ s over a duration of 50 s. The (preliminary data analysis displays for excitation, $F(t)$, and response, $u(t)$, are shown in Figure 4.2.

General uniformity of the spectrogram and time history plots with respect to time for both excitation and response records indicate that the process is *stationary*. Both the spectrogram and autospectrum for the excitation indicate broad band frequency content. The output, $u(t)$, spectrogram and autospectrum reveal narrow band random response character. Gaussian probability density distribution for the response history suggests that the system is a *linear* SDOF system with 20 Hz natural frequency [1].

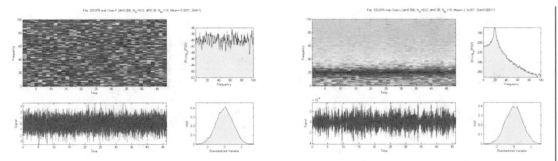

Figure 4.2: Linear SDOF system broadband random excitation (left) and response (right).

## 4.1.10    ILLUSTRATIVE EXAMPLE: ISPE MODAL TEST

In 2017, NASA/MSFC conducted a modal test on the Integrated Spacecraft and Payload Element (ISPE), general overview (courtesy of NASA/MSFC) depicted in Figure 4.3.

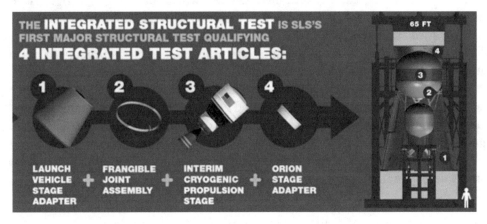

Figure 4.3: ISPE test article overview.

The modal test sensor suite included four electrodynamic shakers (excitation forces) and 273 accelerometer channels (dynamic responses). Preliminary measured data analysis displays for excitation "1" and response "1" are shown in Figure 4.4.

Both the spectrogram and autospectrum for the excitation indicate broadband frequency content. The drop-off in the autospectrum above 90 Hz is the result of low-pass filtering of measured data histories. The response spectrogram and autospectrum reveal the presence of many peaks in the 15–90 Hz frequency band, suggesting the presence of many test article modes. Gaussian probability density distribution for the response history suggests that the ISPE test article exhibited *linear* MDOF system behavior in the 15–90 Hz frequency band.

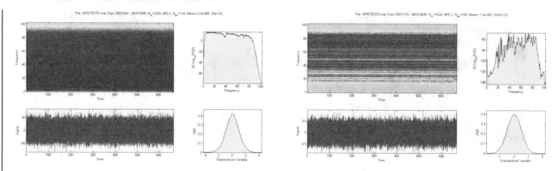

Figure 4.4: ISPE test article broad band random excitation (left) and response (right).

## 4.1.11    ILLUSTRATIVE EXAMPLE: WIRE ROPE TEST ARTICLE

In 2006, during the conducting of U.S. Army sponsored research at Measurement Analysis Corporation, time history data was measured on a test article consisting of a "rigid mass" supported by four wire-rope isolators, which was subjected to random excitation delivered by a vertically oriented electrodynamic shaker. Time histories of acceleration of the rigid mass and the shaker applied load were recorded. Employing a series of mathematical operations on the measured data records, time history records of (a) isolator deflection and (b) isolator internal force were estimated. In one particular test, broadband random excitation was delivered by the shaker. Preliminary data analysis of the estimated isolator deflection and isolator internal force probability density functions is summarized, along with a photograph of the test configuration, in Figure 4.5.

The distortion in isolator internal force probability density with respect to ideal Gaussian probability density indicates the presence of nonlinear behavior [2].

## 4.1.12    ILLUSTRATIVE EXAMPLE: ISS P5 MODAL TEST

During conduct of the ISS P5 modal test in 2001, nonlinear behavior of the test article was suspected. In particular, modal frequencies and mode shapes associated with successive data records were inconsistent with one another. As a result, a series of data records were acquired and subjected to in-depth preliminary data analysis. The source of anomalous dynamic behavior was quickly localized to a left-side trunnion that connected the ISS P5 test article to the test fixture; left- and right-side trunnion locations are illustrated in Figure 4.6.

The preliminary data analysis strategy employed to evaluate and locate suspected nonlinear behavior was based on the theoretical premise that:

a. response of a linear system to sinusoidal excitation at one frequency will also occur at the excitation frequency;

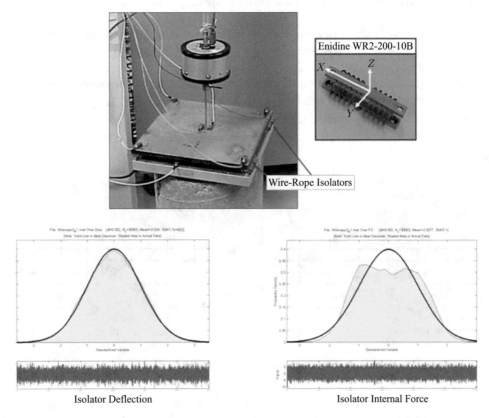

Figure 4.5: Wire rope isolators subjected to broadband random excitation.

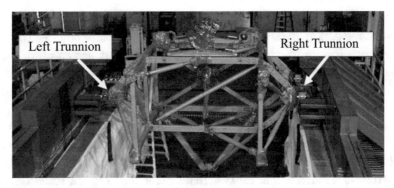

Figure 4.6: ISS P5 test article trunnion locations.

b. response of a system with localized nonlinearity to sinusoidal excitation at one frequency will include frequency components at other frequencies (harmonic distortion); and

c. nonlinear harmonic distortion tends to be most prominent at locations near the source of nonlinear behavior (advice provided by the late Allan G. Piersol).

Employing swept-sine applied load excitation (shaker 2 only), response activity of all 258 measured accelerometer channels was reviewed. Swept-sine responses in the form of spectrograms and focused response spectrum (SRS) calculations were used to identify the source of nonlinear behavior. Results associated with left- and right-side trunnion locations are illustrated in Figure 4.7.

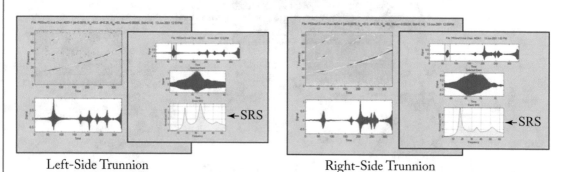

Left-Side Trunnion                          Right-Side Trunnion

Figure 4.7: Trunnion responses to swept-sine excitation.

Clearly, the left-side trunnion indicates strong harmonic distortion (second harmonic) in the computed SRS function, while the right-side trunnion indicates minor harmonic distortion.

## 4.1.13   CLOSURE

Preliminary measured data analysis is vital to integrity of all experimental steps of the integrated test analysis process. Failure to conduct a thorough preliminary screening of measured time history records is known to result in unreliable and ambiguous outcomes for modal and (more general) dynamic characterization tests.

An effective strategy for screening of measured data records includes review of probability densities, autospectra, and spectrograms for all measured time history channels. When the subject test article is subjected to continuous, stationary, broadband random excitation(s) following Gaussian probability distributions, two outcomes are possible: (1) all responses follow Gaussian probability distributions indicating linear system behavior and (2) some responses exhibit probability distributions that deviate from ideal Gaussian distribution indicating nonlinear behavior. Measured ISPE test data indicates clearly linear system behavior, while measured wire rope test data indicates strongly nonlinear behavior.

A second strategy for screening of measured data records includes review of swept sine excitation induced spectrograms for all excitation and response records. The occurrence of harmonic distortion (frequency content other than the excitation frequency) provides an indication of nonlinear dynamic behavior. Moreover, in the case of localized nonlinear behavior, proximity of response channels to a nonlinear mechanism is indicated by severity of harmonic distortion. Application of this strategy was quite effective in localization of nonlinear behavior during ISS P5 modal testing. In addition, focused response spectrum analyses during time segments indicting strongest nonlinear activity provided confirming data related to nonlinear behavior.

## 4.2    PART 2: FREQUENCY RESPONSE FUNCTION ESTIMATES FROM MEASURED DATA

### 4.2.1    INTRODUCTION

The response characteristics of linear dynamic systems, namely frequency response functions (FRFs) are theoretically expressed in the frequency domain based on response to simple harmonic excitation. FRFs are estimated employing spectral analysis techniques [1] of measured excitation and response time history records. Quality of the measured FRF estimates is judged on the basis of coherence functions that are indicative of "signal to noise" ratios as functions of frequency.

Well-developed spectral analysis techniques were adapted in the mid-1990s [2] to address estimation and identification of dynamic systems exhibiting "algebraic" nonlinear behavior. Treatment of systems exhibiting "hysteretic" nonlinear behavior, however generally require employment of specialized time-domain strategies.

After completion of preliminary data evaluations, detailed analyses are often performed to estimate input-output characteristics of a subject dynamic system. The following sections provide an overview of single input/single output (SI/SO) and multiple input/single output (MI/SO) data analysis procedures. Spectral and correlation analysis of SI/SO and MI/SO data records is the primary tool used for estimation of system FRF characteristics.

### 4.2.2    MI/MO FREQUENCY RESPONSE EQUATIONS FOR SYSTEMS WITH ALGEBRAIC NONLINEARITIES

Consider the time domain matrix equation set (introduced in Chapter 3), which includes nonlinear force terms, $F_N(t)$,

$$[M]\{\ddot{U}(t)\} + [B]\{\dot{U}(t)\} + [K]\{U(t)\} = [\Gamma_e]\{F_e(t)\} + [\Gamma_N]\{F_N(t)\}. \tag{4.16}$$

The companion Fourier transform of the above equation set is expressed as

$$\{\ddot{U}(f)\} = \left[ Z(f)^{-1}\Gamma_e \right]\{F_e(f)\} + \left[ Z(f)^{-1}\Gamma_N \right]\{F_N(f)\}$$
$$= [H_e(f)]\{F_e(f)\} + [H_N(f)]\{F_N(f)\}, \tag{4.17}$$

where the frequency domain impedance matrix is defined as

$$[Z(f)] = -(2\pi \cdot f)^{-2}\left[ K + i \cdot (2\pi \cdot f) \cdot B - (2\pi \cdot f)^2 \cdot M \right]. \tag{4.18}$$

The FRF matrices and forces may be "folded" into the all-encompassing matrices and arrays

$$\left[ H_{\ddot{U}F}(f) \right] = \left[ \; H_e(f) \quad H_N(f) \; \right], \qquad \{F(f)\} = \left\{ \begin{array}{c} F_e(f) \\ F_N(f) \end{array} \right\}. \tag{4.19}$$

Therefore, the general FRF relationship that implicitly includes the nonlinear force terms adopts the common linear input-output relationship,

$$\{\ddot{U}(f)\} = \left[ H_{\ddot{U}F}(f) \right]\{F(f)\} \text{ or more generally } \{Y(f)\} = [H_{YX}(f)]\{X(f)\}. \tag{4.20}$$

Benefits associated with "folding" in of the nonlinear partitions into the more convenient linear form will be exploited in development of nonlinear system estimation strategy (for systems with "algebraic" nonlinearities).

### 4.2.3   SI/SO, MI/SO AND MI/MO FREQUENCY RESPONSE FUNCTION ESTIMATION

For simplicity, the frequency domain arrays and FRF matrix described in Equation (4.20) are now presented in a form dropping the brackets and frequency designations. Therefore, the MI/MO relationship describing measured data that includes noise is

$$Y = H \cdot X + N. \tag{4.21}$$

Post-multiplying the above equation set by the complex conjugate transpose (Hermitian conjugate) of the force array results in

$$(Y \cdot X^*) = H \cdot (Y \cdot X^*) + (N \cdot X^*). \tag{4.22}$$

Post-multiplication of Equation (4.21) by its Hermitian conjugate results in

$$(Y \cdot Y^*) = H \cdot (X \cdot X^*) \cdot H^* + (N \cdot X^*) \cdot H^* + H \cdot (N \cdot X^*) + (N \cdot N^*). \tag{4.23}$$

Application of the averaging process used to define the autospectrum and cross-spectrum for individual time history series (see Equations (4.11)–(4.14)), the following spectral matrix relationships are defined (now reintroducing the matrix, array, and frequency labels):

$$[G_{NX}(f)] = [0], \quad [G_{XN}(f)] = [0] \quad \text{(uncorrelated excitations and noise)}, \tag{4.24}$$

$$[G_{YX}(f)] = [H_{YX}(f)] \cdot [G_{XX}(f)] \quad \text{(output-input relationship)}, \tag{4.25}$$

$$[G_{YY}(f)]_C = [H_{XX}(f)] \cdot [G_{XX}(f)] \cdot [H_{YX}(f)]^* \quad \text{(coherent output autospectrum)}, \tag{4.26}$$

$$[G_{YY}(f)] = [G_{YY}(f)]_C + [G_{NN}(f)] \quad \text{(total output autospectrum)}. \tag{4.27}$$

In the case of a single applied force excitation and a single response variable, SI/SO algebraic manipulation yields the following estimates:

$$H_{YX}(f) = G_{YX}(f)/G_{XX}(f) \quad \text{(SI/SO FRF)}, \tag{4.28}$$

$$\gamma_{YX}^2(f) = \frac{|H_{YX}(f)|^2 G_{XX}(f)}{G_{YY}(f)} = \frac{|G_{YX}(f)|^2}{G_{XX}(f) \cdot G_{YY}(f)}$$

$$\leq 1 \quad \text{(SI/SO ordinary coherence)}. \tag{4.29}$$

Multiple Input/Single Output or MI/SO analysis represents a simple extension to SI/SO analysis. The most direct relationship for MI/SO frequency functions involves direct frequency-by-frequency inversion of the cross-spectral matrix of generally partially correlated inputs (forces), i.e.,

$$[H_{YX}(f)] = [G_{YX}(f)] \cdot [G_{XX}(f)]^{-1}, \quad \text{(MI/SO FRF array)}. \tag{4.30}$$

In addition, since the MI/SO coherent output and total output autospectra are non-matrix functions, the total ordinary coherence function is simply,

$$\gamma_{YX}^2(f) = \frac{(G_{YY}(f))_C}{G_{YY}(f)} \leq 1, \quad \text{(MI/SO ordinary coherence)}. \tag{4.31}$$

The above MI/SO formulation does not offer details related to contributions associated with multiple excitations to the output response. An alternative MI/SO analysis formulation using triangular decomposition is thoroughly developed by Bendat and Piersol [1].

Cholesky factorization of the input cross-spectral matrix, $[G_{XX}(f)]$ yields,

$$[G_{XX}(f)] = [\Gamma_{XZ}(f)] \cdot [\Gamma_{ZX}(f)], \tag{4.32}$$

where $[\Gamma_{XZ}(f)]$ is a complex, lower triangular matrix and $[\Gamma_{ZX}(f)]$ is its upper triangular, Hermitian transpose. A triple product decomposition version of Cholesky factors is formed by unit diagonal normalization of $[\Gamma_{XZ}(f)]$ and $[\Gamma_{ZX}(f)]$ resulting in

$$[G_{XX}(f)] = [L_{XZ}(f)] \cdot [G_{ZZ}(f)] \cdot [L_{ZX}(f)], \tag{4.33}$$

where $[L_{XZ}(f)]$ and $[L_{ZX}(f)]$ are lower and upper triangular, unit diagonal matrices, respectively, and $[G_{ZZ}(f)]$ is a positive-definite, diagonal matrix.

Physical significance of the normalized triangular factors is recognized by noting the input variable transformation

$$\{X(f)\} = [L_{XZ}(f)] \cdot \{Z(f)\}. \tag{4.34}$$

The successive terms in $\{Z(f)\}$ are

$$
\begin{aligned}
Z_1(f) &= X_1(f), \\
Z_2(f) &= X_2(f) \quad \text{``swept'' of contributions from } X_1(f), \\
Z_3(f) &= X_3(f) \quad \text{``swept'' of contributions from } X_1(f) \text{ and } X_2(f).\ldots
\end{aligned}
\tag{4.35}
$$

Therefore, $[G_{ZZ}(f)]$ is the diagonal autospectrum matrix of uncorrelated, generalized inputs, $\{Z(f)\}$.

Substitution of the triple product $[G_{XX}(f)]$ decomposition relationship (Equation (4.33)) into Equation (4.30) results in the frequency response functions associated with uncorrelated generalized inputs, $\{Z(f)\}$,

$$[H_{YZ}] = [G_{YZ}] \cdot [G_{ZZ}]^{-1}, \tag{4.36}$$

where

$$[H_{YZ}] = [H_{YX}] \cdot [L_{XZ}], \qquad [G_{YZ}] = [G_{YX}] \cdot [L_{ZX}]^{-1}. \tag{4.37}$$

Therefore, the frequency responses associated with physical inputs, $\{X(f)\}$, are now expressed as

$$[H_{YX}] = [H_{YZ}] \cdot [L_{XZ}]^{-1}. \tag{4.38}$$

The generalized MI/SO equation in terms of generalized inputs is defined as

$$Y(f) = [H_{YZ}(f)] \cdot \{Z(f)\} + N(f). \tag{4.39}$$

Since the generalized inputs, are uncorrelated, the output autospectrum (dropping the "$f$" designation for simplicity) is expressed as

$$G_{YY} = \left|H_{YZ_1}\right|^2 \cdot G_{Z_1 Z_1} + \left|H_{YZ_2}\right|^2 \cdot G_{Z_2 Z_2} + \cdots + \left|H_{YZ_N}\right|^2 \cdot G_{Z_K Z_K} + G_{NN}. \tag{4.40}$$

Coherent output autospectra and *cumulative coherence function* pairs are therefore defined as follows:

$$(G_{YY})_1 = \left|H_{YZ_1}\right|^2 \cdot G_{Z_1 Z_1}, \quad \gamma_{Y1}^2 = \left(\left|H_{YZ_1}\right|^2 \cdot G_{Z_1 Z_1}\right)/G_{YY} \text{ for input } x_1(t), \tag{4.41}$$

$$(G_{YY})_2 = \left|H_{YZ_1}\right|^2 \cdot G_{Z_1 Z_1} + \left|H_{YZ_2}\right|^2 \cdot G_{Z_2 Z_2},$$

$$\gamma_{Y2}^2 = \left(\left|H_{YZ_1}\right|^2 \cdot G_{Z_1 Z_1} + \left|H_{YZ_2}\right|^2 \cdot G_{Z_2 Z_2}\right)/G_{YY} \text{ for inputs } x_1(t) + x_2(t), \tag{4.42}$$

$$(G_{YY})_K = \left|H_{YZ_1}\right|^2 \cdot G_{Z_1 Z_1} + \left|H_{YZ_2}\right|^2 \cdot G_{Z_2 Z_2} + \cdots + \left|H_{YZ_N}\right|^2 \cdot G_{Z_K Z_K},$$

$$\gamma_{YK}^2 = \left(\left|H_{YZ_1}\right|^2 \cdot G_{Z_1 Z_1} + \left|H_{YZ_2}\right|^2 \cdot G_{Z_2 Z_2} + \cdots + \left|H_{YZ_N}\right|^2 \cdot G_{Z_K Z_K}\right)/G_{YY},$$

$$\text{for inputs, } x_1(t) + \cdots + x_k(t). \tag{4.43}$$

The cumulative coherence function family has the property

$$0 \le \gamma_{Y1}^2 \le \gamma_{Y2}^2 \le \cdots \le \gamma_{YK}^2 \le 1. \tag{4.44}$$

When graphically displayed as a function of frequency, the coherence function family appears as a waterfall plot series that clearly indicates the relative contributions of the accumulated excitation sources. The benefits of such a display will be demonstrated in a series of illustrative examples.

Finally, it is noted that MI/MO analysis represents a simple extension to MI/SO analysis.

## 4.2.4 ILLUSTRATIVE EXAMPLE: ISPE MODAL TEST

The ISPE test article was excited with broad band random excitation forces at four separate locations, and MI/MO correlation and spectral analysis was preformed employing 8192 length windows with 50% overlap processing. Cumulative coherences for all 265 TAM response channels plus 4 drive point response channels were computed. MI/MO plots associated with all 4 excitation forces and 4 drive point responses are detailed in Figure 4.8.

The plots in Figure 4.8 are arranged in a 4 × 4 matrix "plot map." For example, the plots in the (2,3) position correspond to excitation 2, drive point response 3. Each plot "$i, j$" entry includes the cumulative coherence (top), FRF phase angle (middle), and FRF magnitude (bottom). An important feature of the plot matrix is FRF "reciprocity" (e.g., the "$i, j$" component is consistent with the "$j, i$" component, with regard to both magnitude and phase of the FRF). Greater detail can be discerned from the plot group for the "1, 1" component shown in Figure 4.9. Of particular interest is the (top) cumulative coherence plot, which (a) indicates the successive contributions of the first three excitations (to drive point 1 response autospectum) that are substantially below unity and (b) indicates the significant role of the fourth excitation producing near unit cumulative coherence.

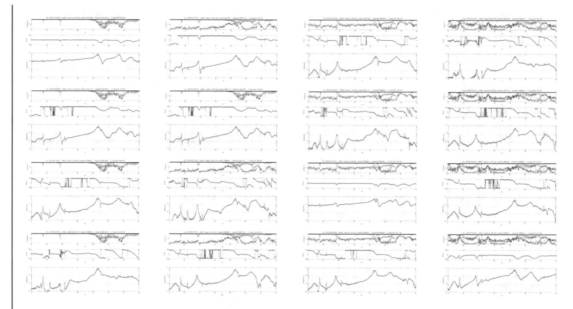

Figure 4.8: Representative ISPE MI/MO FRF and cumulative coherence plots.

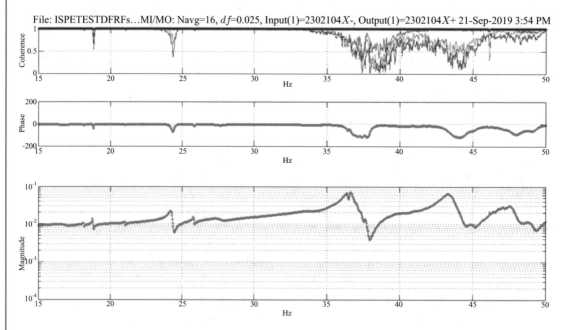

Figure 4.9: ISPE MI/MO FRF and cumulative coherence plots for excitation 1, drive point response 1.

The cumulative coherences associated with nearly all response channels were close to unity confirming extremely high-quality FRF estimates.

### 4.2.5    ILLUSTRATIVE EXAMPLE: WIRE ROPE MI/SO TEST DATA ANALYSIS

Recalling the preliminary measured data analysis, summarized in Figure 4.5, the hypothesized "algebraic" nonlinear system composed of measured time histories (applied force and acceleration response) and synthesized "measured" time histories (cubed displacement and velocity·|velocity|) depicted in Figure 4.10 was subjected to MI/SO analysis.

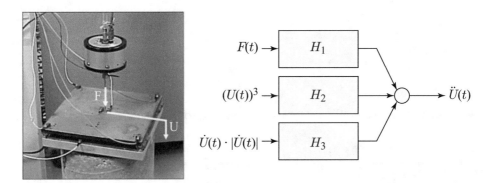

Figure 4.10: Hypothesized MI/SO nonlinear system.

The cumulative coherence plot, shown in Figure 4.11 indicates that incorporation of the two nonlinear terms produces a nearly unit value cumulative coherence (red curve), while the ordinary coherence associated with a linear model (blue curve) indicates reduced coherence.

The results from Figure 4.11 provide clear evidence that the behavior of the wire rope isolators is nonlinear. However, further definitive data analysis, described in Chapter 7, will indicate that the nonlinear behavior is "hysteretic" rather than "algebraic" as currently hypothesized. That being said, the present "algebraic" nonlinear model serves an important role as an "intermediate" data analysis.

### 4.2.6    ILLUSTRATIVE EXAMPLE: ISS P5 MODAL TEST

The ISS P5 test article was excited with broadband random excitation forces at three separate locations, MI/MO correlation and spectral analysis was preformed employing 1024 length windows with 50% overlap processing. Cumulative coherences for all 261 TAM response channels plus 3 drive point response channels were computed. MI/MO plots associated with the three drive point FRFs (complete drive point FRF "matrix" as depicted for the ISPE in Figure 4.8 is not provided here) are detailed in Figure 4.12.

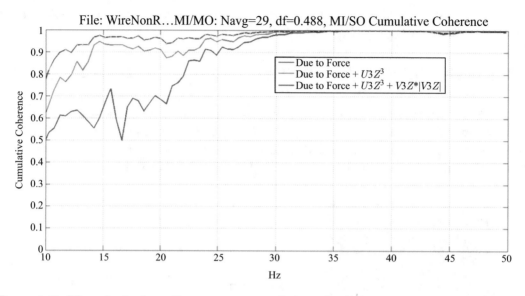

Figure 4.11: Hypothesized nonlinear system cumulative coherence.

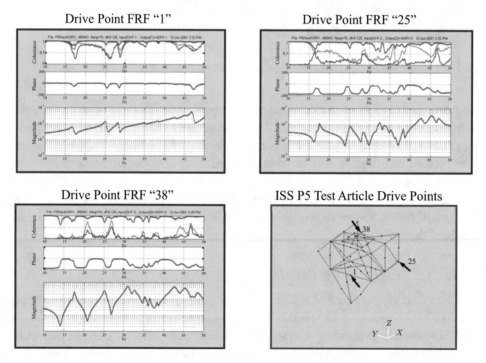

Figure 4.12: ISS P5 MI/MO FRF and cumulative coherence plots.

The reduced cumulative coherences in the 30–35 Hz frequency band were judged provide additional evidence of test article nonlinear behavior, which was noted in preliminary data analysis results summarized in Figure 4.7.

### 4.2.7  CLOSURE

Systematic techniques for estimation of FRFs associated with single and multiple excitation resources, due to Bendat and Piersol, provide a sound foundation for detailed measured analysis of measured time history records. Automation and reinterpretation of triangular (Cholesky) decomposition of the excitation cross-spectral (frequency dependent) matrix leads to a straightforward, unambiguous cumulative coherence function family. The cumulative coherence family provides insight into the role played by each individual excitation (which is not necessarily completely uncorrelated with the other excitation sources). Benefits offered by cumulative coherence, as demonstrated with modal test data records, include: (1) further confirmation and localization of ISS P5 test article behavior (noted in preliminary data analyses) and (2) confirmation of ISPE test article linear behavior, along with determination of the relative prominence of modal test excitation resources. In addition, employment of cumulative coherence facilitates characterization of "algebraic" nonlinear systems as demonstrated in testing of a wire rope shock and vibration isolator.

## 4.3  REFERENCES

[1] J. S. Bendat and A. G. Piersol, *Random Data Analysis and Measurement Procedures*, 4th ed., Wiley, 2010. DOI: 10.1002/9781118032428. 65, 66, 67, 68, 69, 70, 75, 77

[2] J. S. Bendat, *Nonlinear Systems Techniques and Applications*, 2nd ed., Wiley, 1998. 70, 72, 75

# CHAPTER 5

# Experimental Modal Analysis

## 5.1    PART 1: PRELIMINARY EXPERIMENTAL MODAL ANALYSIS

### 5.1.1    INTRODUCTION

The discipline of experimental modal analysis developed over two distinct time periods, namely: (1) the analog era prior to 1971 and (2) the digital era from 1971 to the present time. Key publications offering an historical perspective on the discipline have been authored by Bishop and Gladwell [1], covering the analog era up to 1961, and Brown and Allemang [2], spanning the analog and digital eras through 2007. In spite of availability of many excellent automated experimental modal analysis procedures, developed during the digital era, review of measured FRF data (preliminary experimental modal analysis) is always a prudent step in the overall experimental modal analysis process. Preliminary experimental modal analysis is highly dependent on analysis of graphical FRF displays based on (1) strategies dating back to the analog era and (2) modal indicator displays introduced during the digital era [3].

### 5.1.2    FREQUENCY RESPONSE CHARACTERISTICS OF LINEAR SDOF SYSTEMS

An elementary linear structural dynamic or mechanical dynamic system is described in terms of a SDOF linear mechanical system. The dynamic response, $u(t)$, of a SDOF system which is excited by an applied force, $F(t)$ and/or base (foundation) motion input, $u_0(t)$, is governed by the ordinary differential equation (schematic depicted in Figure 5.1),

$$m\ddot{u}(t) + b\dot{u}(t) + ku(t) = F(t) + b\dot{u}_0(t) + ku_0(t), \tag{5.1}$$

where $m$, $b$, and $k$ are the constant mass, viscous damping, and elastic stiffness coefficients, respectively. By defining the relative displacement variable, $u_R(t) = u(t) - u_0(t)$, and dividing by the mass, $m$, Equation (2.1) simplifies to

$$\ddot{u}_R(t) + 2\zeta_n \omega_n \dot{u}_R(t) + \omega_n^2 u_R(t) = F(t)/m - \ddot{u}_0(t), \tag{5.2}$$

where the undamped natural frequency (rad/sec) and critical damping ratio, respectively, are,

$$\omega_n = \sqrt{\frac{k}{m}}, \qquad \zeta_n = \frac{b}{(2m\omega_n)}. \tag{5.3}$$

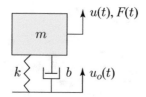

Figure 5.1: Linear SDOF dynamic system schematic.

In addition, the damped natural frequency (rad/sec) is defined as,

$$\omega_d = \omega_n \sqrt{1 - \zeta_n^2}. \tag{5.4}$$

Steady-state response of a SDOF system subjected to simple harmonic force excitation, $F(\omega)e^{i\omega t}$, is of the form, $U(\omega)e^{i\omega t}$, where the following FRF quantities are derived after substitutions into the system differential equation (and conversion of the frequency variable, $\omega = 2\pi f$):

$$H(f) = \frac{ku(f)}{F(f)} = \frac{1}{(1 - (f/f_n)^2 + 2i\zeta_n(f/f_n))} \quad \text{(displacement FRF)}, \tag{5.5}$$

$$H_A(f) = \frac{m\ddot{u}(f)}{F(f)} = \frac{-(f/f_n)^2}{(1 - (f/f_n)^2 + 2i\zeta_n(f/f_n))} \quad \text{(acceleration FRF)}, \tag{5.6}$$

$$TR(f) = \frac{b\dot{u}(f) + ku(f)}{F(f)} = \frac{(1 + 2i\zeta_n(f/f_n))}{(1 - (f/f_n)^2 + 2i\zeta_n(f/f_n))} \quad \text{(transmissibility)}. \tag{5.7}$$

Steady-state response of a SDOF system subjected to harmonic base motion excitation, $u_0(\omega)e^{i\omega t}$, is of the form $u_R(\omega)e^{i\omega t}$, where the following FRF quantities are derived after substitutions into the system differential equation:

$$H_A(f) = \frac{-\ddot{u}_R(f)}{\ddot{u}_0(f)} = \frac{-(f/f_n)^2}{(1 - (f/f_n)^2 + 2i\zeta_n(f/f_n))} \quad \text{(relative acceleration FRF)}, \tag{5.8}$$

$$TR(f) = \frac{\ddot{u}(f)}{\ddot{u}_0(f)} = \frac{(1 + 2i\zeta_n(f/f_n))}{(1 - (f/f_n)^2 + 2i\zeta_n(f/f_n))} \quad \text{(transmissibility)}. \tag{5.9}$$

Each of the above frequency response functions exhibits peak response amplification when the excitation occurs near the natural frequency. Moreover, the phase of each FRF shifts 180° as the excitation frequency increases from below to above the natural frequency. Shown in Figures 5.2–5.4 are plots of the three types of FRFs in terms of magnitude and phase, real and imaginary parts vs. frequency ($f/f_n$), and real vs. imaginary parts (where $f_n = 1, \zeta_n = .02$).

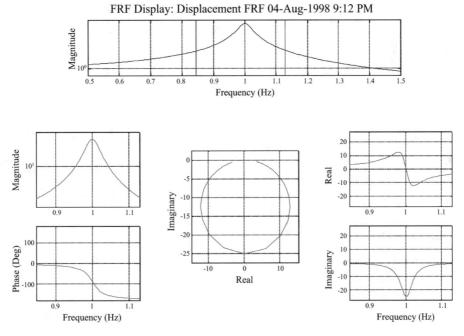

Figure 5.2: SDOF displacement FRF, $H(f)$ vs. $f$.

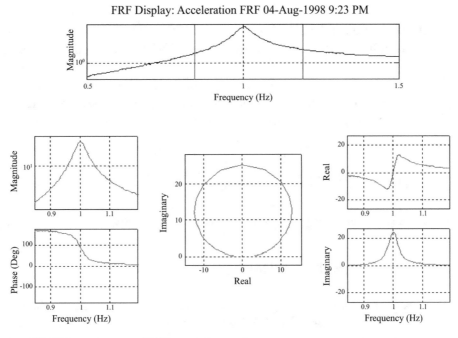

Figure 5.3: SDOF acceleration FRF, $H_a(f)$ vs. $f$.

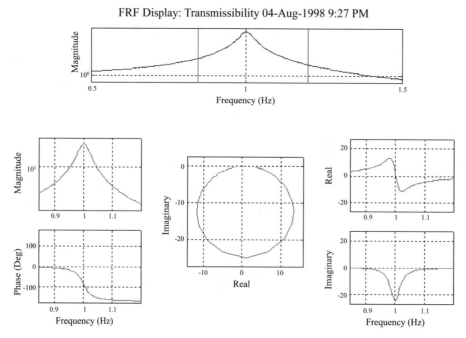

Figure 5.4: SDOF transmissibility, $TR(f)$ vs. $f$.

A closer examination of the SDOF acceleration FRF reveals some well-known properties of this (as well as SDOF displacement and transmissibility FRF) response functions, as displayed in Figure 5.5.

In particular, the red dot near the peak value of the FRF magnitude, imaginary component, and polar plot is indicative of the natural frequency, $f_n$. In addition, the frequency spacing of peaks in the imaginary component, indicated by the green dots, facilitates estimation of damping.

### 5.1.3   FREQUENCY RESPONSE CHARACTERISTICS OF LINEAR MDOF SYSTEMS

Consider a structural dynamic system subjected to applied excitation forces,

$$[M]\{\ddot{u}\} + [B]\{\dot{u}\} + [K]\{u\} = [\Gamma_e]\{F_e\}. \tag{5.10}$$

Employing the assumption of simple harmonic excitation and responses, in the same manner as employed for SDOF systems, MDOF frequency response to each separate "unit" excitation force is described as

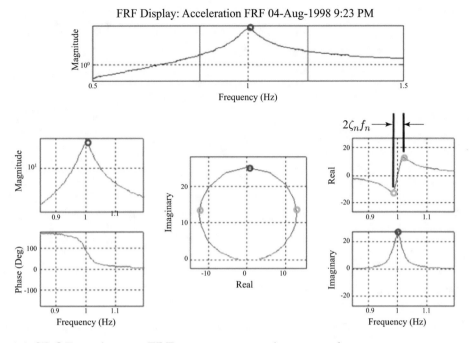

Figure 5.5: SDOF acceleration FRF, $H_a(f)$ vs. $f$ with annotated points.

$$\{H(\omega)\} = \{\ddot{U}(\omega)\} = -\omega^2 \left[ K + i\omega[B] - \omega^2 M \right]^{-1} \{\Gamma_e\} \quad \text{(noting that } \omega = 2\pi f). \quad (5.11)$$

Estimation of MDOF system frequency responses from measured time history data was described in Chapter 4 along with illustrative example processed test data. It is instructive at this point to review some of the ISPE modal test estimated FRF functions within the context of preliminary experimental modal analysis.

### 5.1.4    ILLUSTRATIVE EXAMPLE: ISS P5 MODAL TEST

While preliminary experimental modal analysis (employing review of measured FRF data) was conducted during P5 modal testing at NASA MSFC, the primary focus of that effort shifted toward preliminary investigation of nonlinear behavior (as described in Chapter 4) and non-repeatability (sensitivity) of the test article dynamic characteristics. Therefore, further discussion of the ISS P5 modal test will resume in Part 2 of this chapter.

### 5.1.5    ILLUSTRATIVE EXAMPLE: ISPE MODAL TEST

Experimental FRF displays associated with drive point 1, and two selected response DOFs are provided in Figures 5.6 and 5.7, respectively.

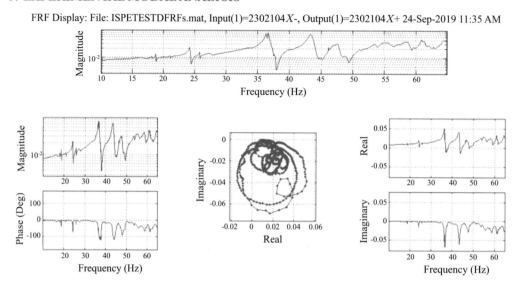

Figure 5.6: ISPE drive point 1 FRF detail.

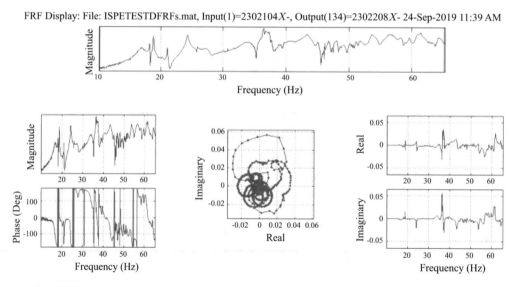

Figure 5.7: ISPE response point FRF detail.

The above two FRFs are certainly more complicated than the idealized SDOF FRFs depicted in Figures 5.2–5.5.

A more comprehensive overview of the "many modes" evident in the ISPE test article is evidenced by display of two composite signature functions, namely: (a) a normalized square-root of the summation of the autospectra for all modal test response accelerations and (b) a global FRF skyline function. The global FRF skyline function is defined as a complex frequency domain function whose real and imaginary parts are the sum of the absolute values of the real component of response FRFs and the sum of the absolute values of the imaginary component of response FRFs, that is,

$$SKY(f) = \sum_{k=1}^{N_H} |real\,(H_k\,(f))| + i \sum_{k=1}^{N_H} |imag\,(H_k\,(f))|\,. \tag{5.12}$$

A display of the two global skyline functions for the ISPE test article is provided in Figure 5.8.

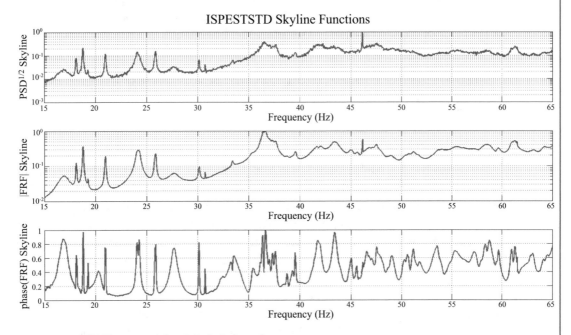

Figure 5.8: ISPE test article global skyline functions.

The phase FRF skyline offers the clearest indication of the presence of many modes in the 15–65 Hz frequency band. While the skylines indicate the presence of "many modes," further investigation of local response FRFs will provide insight into the presence of additional localized modes (a difficult, slow task for a structural test article that is mapped by 265 response channels and 4 excitations, that is 1060 individual FRFs!). Other strategies for discerning the presence of "many modes" as well as closely spaced modes have been developed and successfully applied by leaders in the experimental structural dynamics community [3].

## 5.1.6   CLOSURE

Several intuitive experimental modal analysis tools have been described. SDOF frequency response functions displayed in three ways, namely: (1) magnitude and phase components vs. frequency; (2) real and imaginary components vs. frequency (also called co- and quad- functions); and (3) polar real vs. imaginary components vs. frequency. All three display types provide intuitive, well-known means for estimation of modal frequency and damping parameters. Review of FRFs associated with individual FRFs for actual MDOF systems offer some insights for estimation of modal frequency and damping parameters when modes are well separated (not the case for the ISPE test article and other shell-type structures).

Global skyline functions offer some insight into the presence of multiple modes in a designated frequency band. However, such functions do not necessarily highlight the presence of localized modes. A more comprehensive approach involving review of many or all measured FRF response functions is a tedious, time-consuming endeavor.

# 5.2   PART 2: THE SIMULTANEOUS FREQUENCY DOMAIN METHOD

## 5.2.1   INTRODUCTION

Difficulties encountered by NASA/MSFC on the Integrated Spacecraft Payload Element (ISPE) modal survey in the fall of 2016 bring an important challenge to the forefront. Specifically, which estimated test modes are "authentic," and which modes are due to "noise" associated with measured FRFs? The present discussion on experimental modal analysis (EMA) focuses on mathematical isolation of individual estimated test mode FRFs in a manner that is similar to the concept developed by Mayes and Klenke [4]. While the presently discussed EMA approach ought to be quite independent of the investigator's choice of experimental modal analysis algorithm, the results herein apply to methods that explicitly estimate the tested system's state-space plant matrix such as the Simultaneous Frequency Domain (SFD) method [5–7].

The latest version of SFD (SFD-2018) employs mathematical operations aimed at isolating individual candidate experimental modes without direct reliance on information associated with the subject system's TAM) mass matrix. The key to mathematical and visual isolation of individual modes from measured data is the left-hand eigenvector. Virtually all modern experimental modal analysis techniques produce estimates of (right-hand) eigenvectors and eigenvalues (modal frequency and damping). While techniques for estimation of left-hand eigenvectors are well-known (e.g., the "real mode transpose times TAM mass matrix product" and the Moore–Penrose pseudo-inverse [8]), they have been judged inadequate. The purest approach to estimation of left-hand eigenvectors is a consequence specifically those techniques that estimate the measured system's plant or effective dynamic system matrix, such as SFD. Since a complete set of raw experimental modes are identified consistent with the order of the estimated plant, the left-hand eigenvectors are calculated exactly from the inverse of the complete, raw right-hand eigenvector set.

## 5.2.2   EFFECTIVE DYNAMIC SYSTEMS

The SFD method [5], introduced in 1981, has undergone substantial revision and refinement since that time [6, 7], primarily by this writer and principals at The Aerospace Corporation. SFD implicitly assumes that FRFs associated with a series of "$N$" excitations may be described in terms of a transformation described by

$$\left[\ \ddot{U}_1(f)\quad \ddot{U}_2(f)\quad \ldots\quad \ddot{U}_N(f)\ \right] = [V]\left[\ \ddot{\xi}_1(f)\quad \ddot{\xi}_2(f)\quad \ldots\quad \ddot{\xi}_N(f)\ \right]. \tag{5.13}$$

By performing singular value decomposition (SVD) analysis [9] of the FRF collection, a dominant set of real generalized trial vectors, $[V]$, and complex generalized FRFs, $\left[\ \ddot{\xi}_1(f)\quad \ddot{\xi}_2(f)\quad \ldots\quad \ddot{\xi}_N(f)\ \right]$, is obtained. Unit normalization of the SVD trial vectors is expressed as

$$[V]^T\,[V] = [I]. \tag{5.14}$$

Theoretically, the generalized FRF array describes the following dynamic system equations associated with the individual applied forces

$$\left[\ddot{\xi}(f)\right] + \left[\tilde{B}\right]\left[\dot{\xi}(f)\right] + \left[\tilde{K}\right][\xi(f)] = \left[\tilde{\Gamma}\right][F(f)],$$
$$\text{where } \left[\tilde{B}\right] = \left[M^{-1}\right][B], \quad \left[\tilde{K}\right] = \left[M^{-1}\right][K],$$
$$\text{and } \left[\tilde{\Gamma}\right] = \left[M^{-1}\right][\Gamma]. \tag{5.15}$$

The real, effective dynamic system matrices, $\left[\tilde{B}\right]$, $\left[\tilde{K}\right]$, and $\left[\tilde{\Gamma}\right]$, are estimated by linear least-squares analysis [10].

Estimation of experimental modal parameters is performed by complex eigenvalue analysis of the state variable form of the effective dynamic system,

$$\left\{\begin{matrix} \ddot{\xi} \\ \dot{\xi} \end{matrix}\right\} = \left[\begin{matrix} -\tilde{B} & -\tilde{K} \\ I & 0 \end{matrix}\right]\left\{\begin{matrix} \dot{\xi} \\ \xi \end{matrix}\right\} + \left[\begin{matrix} \tilde{\Gamma} \\ 0 \end{matrix}\right]\{F\}, \tag{5.16}$$

which is of the general type

$$\{\dot{\eta}\} = \left[A_\eta\right]\{\eta\} + \left[\Gamma_\eta\right]\{F\}. \tag{5.17}$$

Complex eigenvalue analysis of the effective dynamic system produces the following results:

(a) $\{\eta\} = [\Phi_\eta] \{q\}$, where the "left-handed" eigenvectors are $[\Phi_{\eta L}] = [\Phi_\eta]^{-1}$,

(b) $[\Phi_{\eta L}] \cdot [\Phi_\eta] = [I]$,      $[\Phi_{\eta L}] \cdot [A_\eta][\Phi_\eta] = [\lambda]$ (complex eigenvalues),

(c) $[\Phi_{\eta L}] \cdot [\Gamma_\eta] = [\gamma]$ (modal gains), and

(d) $\dot{q}_j - \lambda_j q_j = (\gamma)_j [F(f)]$ (frequency response of individual modes).

$$(5.18)$$

Recovery of experimental modes in terms of the physical DOFs involves back transformation employing the trial vector matrix, $[V]$, specifically,

$$[\Phi] = [V][\Phi_\eta], \qquad [\Phi_L] = [\Phi_{\eta L}][V^T], \qquad [OR] = [\Phi_L][\Phi] \equiv [I]. \qquad (5.19)$$

### 5.2.3   THE SFD METHOD PRIOR TO 2018

Estimation of the effective dynamic system with the SFD method (and more generally any method that performs similar system "plant" estimation operations) will pick up spurious "noise" degrees of freedom and associated spurious modes. Over the years since 1981, the writer has employed a heuristic practice in versions of SFD algorithms that select "authentic" modes from the complete set, which is estimated in selected frequency bands. The heuristic criteria include: (1) elimination of modes having negative damping; (2) modes with very low modal gain; and (3) other modes that appear spurious from any number of physical/experience based considerations. Prior to 2018, the SFD method (this writer's version) did not make use of the complex modes associated with the effective dynamic system (Equations (5.15)–(5.19)).

The theoretical relationship between FRFs and modal parameters (assuming that modal vectors are real) is

$$[H(f)] = [\Phi] \cdot [h(f)], \qquad (5.20)$$

where $[\Phi]$ is the unknown real modal matrix and $[h(f)]$ is the SDOF acceleration FRF matrix. The terms of $[h(f)]$ are defined as

$$h_n(f) = \frac{-(f/f_n)^2}{(1 + 2i\zeta_n(f/f_n) - (f/f_n)^2)}, \qquad (5.21)$$

where the $f_n$ and $\zeta_n$ are the modal frequency and damping associated with the particular experimental mode. Since the modal SDOF acceleration matrix is completely known, the real modal matrix is obtained by linear least squares analysis. At the experimental modal analyst's discretion (highly recommended), low- and/or high-frequency residual modal frequencies may be added to the set of identified eigenvalues (the low-frequency residual FRF has a frequency close to "0" and a user-selected damping value, e.g., $\zeta_n = .01$, and the high-frequency residual has a frequency substantially higher than the highest experimental modal frequency and a user-selected damping value, e.g., $\zeta_n = .01$) to enhance accuracy of modal vector estimates.

The theoretical relationship between FRFs and modal parameters (assuming that modal vectors are complex) is

$$[H(f)] = -(2\pi \cdot f) \cdot \sum_{n=1}^{N} \left( \frac{\Phi_n}{\lambda_n - i(2\pi \cdot f)} + \frac{\Phi_n^*}{\lambda_n^* - i(2\pi \cdot f)} \right), \qquad (5.22)$$

where $\Phi_n$ is the unknown $n$th complex modal residue vector and $\lambda_n$ is the $n$th (positive imaginary part) complex eigenvalue. The complex modal residue vectors are proportional to the complex system modes. Complex eigenvalues (when critical damping ratio, $\zeta_J$, is less than 1.0) are

$$\lambda_n = -\zeta_n \omega_n + i\omega_n. \qquad (5.23)$$

Since the complex eigenvalues are completely known, the complex modal residue vectors (proportional to complex modal vectors) are obtained by linear least squares analysis. As for the case of real modes, low and/or high frequency residual modal frequencies may be added to the set of identified eigenvalues to enhance accuracy of modal vector estimates.

Regardless of whether the user chooses to estimate real or complex modal vectors, reconstructed FRFs calculated from identified modal parameters serve as a means for a quality check on the overall (pre-2018) SFD process.

## 5.2.4   ILLUSTRATIVE EXAMPLE: ISS P5 MODAL TEST

Due to problematic nonlinear behavior and variability of modal characteristics of the ISS P5 test article, 17 separate measured time tistory data sets were recorded and evaluated employing the preliminary and detailed data analysis techniques described in Chapter 4. An overview of the measured data sets is provided in Table 5.1.

The three highlighted sets were selected by NASA and the test team for detailed experimental data analysis, while all measured data sets were employed for the nonlinear-sensitivity investigation.

Detailed experimental modal analysis of measured data set TSS2, employing the (pre-2018) SFD method, resulted in estimation of 23 test article modes in the 0–63 Hz frequency band as summarized in Table 5.2. Note the first 10 modes (shaded in yellow) were selected for subsequent test-analysis correlation and model reconciliation analyses.

Integrity of the 23 estimated ISS P5 modes for measured data set TSS2 was established by comparison of MI/MO and reconstructed FRFs calculated from identified modal parameters as illustrated in Figure 5.9.

Table 5.1: ISS P5 measured data sets

| Data Set | Duration, Averages | Excitation Type | Excitation Level, Shakers |
|----------|--------------------|-----------------|----------------------------|
| TSS1 | 13.3 minutes, 100 averages | Burst random | 10 lb RMS, 3 shakers |
| **TSS2** | **10 minutes, 75 averages** | **Continuous Random** | **10 lb RMS, 3 shakers** |
| TSS3 | 13.3 minutes, 100 averages | Burst random | 10 lb RMS, shakers 1&2 |
| **TSS4** | **10 minutes, 75 averages** | **Continuous Random** | **10 lb RMS, shakers 1&2** |
| TSS5 | 13.3 minutes, 100 averages | Burst random | 10 lb RMS, shakers 1&2 |
| TSS6 | 13.3 minutes, 100 averages | Burst random | 10 lb RMS, 3 shakers |
| TSS7 | 13.3 minutes, 100 averages | Burst random | 25 lb RMS, 3 shakers |
| TSS8 | 10 minutes, 75 averages | Continuous Random | 25 lb RMS, 3 shakers |
| TSS9 | "long," max overlap | Sine Sweep Up-Down | 3 lb up-10 lb down, shakers 1 |
| TSS10 | "long," max overlap | Sine Sweep Up-Down | 3 lb up-10 lb down, shakers 2 |
| TSS11 | "long," max overlap | Sine Sweep Up-Down | 3 lb up-10 lb down, shakers 3 |
| TSS12 | 6.7 minutes, max overlap | Sine Sweep Up | 10 lb, shaker 3 |
| TSS13 | 6.7 minutes, max overlap | Sine Sweep Up | 10 lb, shaker 2 |
| TSS14 | 6.7 minutes, max overlap | Sine Sweep Up | 10 lb, shaker 1 |
| TSS15 | 6.7 minutes, 50 averages | Burst random | 10 lb RMS, 3 shakers |
| TSS16 | 13.3 minutes, 100 averages | Burst random | 10 lb RMS, 3 shakers |
| **TSS17** | **10 minutes, 75 averages** | **Continuous Random** | **10 lb RMS, 3 shakers** |

Note that the MI/MO estimated FRF is indicated by the blue dots, the reconstructed FRF is indicated by the green curves, and the frequencies associated with the 23 estimated modes are indicated by the red circles.

Following established NASA and USAF Space Command practices [11, 12], orthogonality of the 23 ISS P5 (data set TSS2) experimental modes, expressed in percentage units) is summarized in Figure 5.10 and Table 5.3, respectively.

Review of the contents of Table 5.3 indicates that the NASA orthogonality criterion (goal) [11] was achieved for most of the 23 test modes, including modes 11–23 which had non-negligible kinetic energy content associated with the test fixture.

Table 5.2: **ISS P5 estimated modes for data set TSS2**

| Mode | Freq (Hz) | Zeta (%) | Directional $KE$ (%) | | | Component $KE$ (%) | | |
|---|---|---|---|---|---|---|---|---|
| | | | $X$ | $Y$ | $Z$ | Truss | Grapple | Fixture |
| 1 | 16.94 | 2.13 | 73 | 27 | 1 | 85 | 15 | 0 |
| 2 | 17.58 | 1.08 | 24 | 73 | 2 | 91 | 9 | 0 |
| 3 | 25.19 | 1.08 | 68 | 4 | 28 | 77 | 23 | 0 |
| 4 | 28.44 | 0.80 | 48 | 47 | 5 | 94 | 6 | 0 |
| 5 | 31.12 | 1.25 | 72 | 8 | 20 | 50 | 50 | 0 |
| 6 | 32.6 | 0.84 | 18 | 17 | 65 | 61 | 39 | 0 |
| 7 | 33.66 | 1.55 | 15 | 11 | 74 | 54 | 46 | 0 |
| 8 | 35.19 | 0.72 | 33 | 60 | 7 | 95 | 5 | 0 |
| 9 | 36.39 | 0.40 | 81 | 11 | 8 | 98 | 2 | 0 |
| 10 | 38.38 | 0.32 | 4 | 29 | 67 | 99 | 1 | 0 |
| 11 | 42.86 | 1.44 | 13 | 34 | 53 | 68 | 31 | 1 |
| 12 | 45.16 | 1.49 | 25 | 36 | 39 | 78 | 22 | 0 |
| 13 | 46.95 | 0.79 | 9 | 18 | 74 | 82 | 10 | 8 |
| 14 | 48.4 | 1.04 | 40 | 17 | 43 | 69 | 24 | 7 |
| 15 | 49.1 | 2.30 | 34 | 45 | 22 | 53 | 5 | 42 |
| 16 | 51.69 | 1.15 | 19 | 14 | 67 | 60 | 34 | 6 |
| 17 | 54.06 | 1.54 | 22 | 43 | 35 | 70 | 21 | 9 |
| 18 | 54.45 | 0.83 | 65 | 25 | 10 | 95 | 3 | 2 |
| 19 | 55.3 | 0.67 | 45 | 43 | 12 | 98 | 2 | 1 |
| 20 | 58.1 | 0.84 | 71 | 16 | 13 | 97 | 2 | 0 |
| 21 | 60.15 | 1.55 | 9 | 17 | 74 | 81 | 12 | 7 |
| 22 | 61.78 | 0.04 | 1 | 96 | 4 | 3 | 0 | 97 |
| 23 | 62.84 | 0.65 | 16 | 8 | 76 | 25 | 72 | 3 |

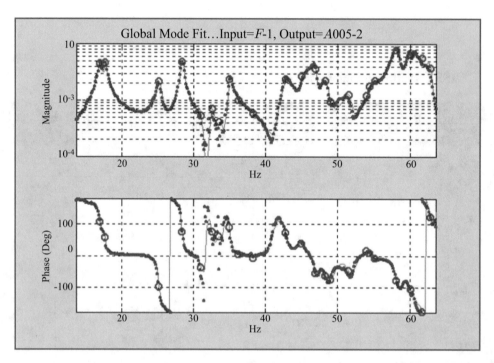

Figure 5.9: Typical ISS P5 MI/MO FRF reconstruction based on pre-2018 SFD methodology.

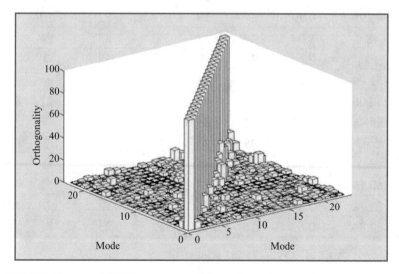

Figure 5.10: ISS P5 (data set TSS2) experimental mode orthogonality.

Table 5.3: ISS P5 (data set TSS2) experimental mode orthogonality

| Mode | Freq (Hz) | Orthogonality (%) | | | | | | | | | | | | | | | | | | | | | | |
|---|---|---|---|---|---|---|---|---|---|---|---|---|---|---|---|---|---|---|---|---|---|---|---|---|---|
| 1 | 16.94 | 100 | 7 | 0 | 0 | 2 | 1 | 2 | 3 | 2 | 2 | 0 | 1 | 0 | 1 | 1 | 0 | 4 | 1 | 0 | 0 | 2 | 0 | 0 |
| 2 | 17.58 | 7 | 100 | 3 | 0 | 3 | 0 | 1 | 3 | 1 | 1 | 1 | 2 | 1 | 2 | 2 | 2 | 1 | 0 | 1 | 3 | 1 | 0 | 1 |
| 3 | 25.19 | 0 | 3 | 100 | 3 | 3 | 1 | 0 | 1 | 1 | 1 | 0 | 1 | 1 | 3 | 3 | 1 | 3 | 0 | 1 | 3 | 3 | 0 | 3 |
| 4 | 28.44 | 0 | 0 | 3 | 100 | 0 | 2 | 3 | 6 | 2 | 0 | 0 | 2 | 1 | 1 | 1 | 1 | 0 | 1 | 0 | 6 | 2 | 0 | 1 |
| 5 | 31.12 | 2 | 3 | 3 | 0 | 100 | 6 | 0 | 2 | 2 | 2 | 0 | 2 | 3 | 6 | 2 | 2 | 1 | 1 | 2 | 2 | 3 | 1 | 0 |
| 6 | 32.60 | 1 | 0 | 1 | 2 | 6 | 100 | 14 | 1 | 2 | 3 | 2 | 3 | 4 | 2 | 1 | 3 | 3 | 2 | 0 | 0 | 2 | 1 | 5 |
| 7 | 33.66 | 2 | 1 | 0 | 3 | 0 | 14 | 100 | 8 | 2 | 1 | 6 | 1 | 0 | 3 | 1 | 1 | 0 | 0 | 2 | 0 | 0 | 1 | 4 |
| 8 | 35.19 | 3 | 3 | 1 | 6 | 2 | 1 | 8 | 100 | 4 | 5 | 3 | 1 | 3 | 0 | 1 | 2 | 1 | 1 | 1 | 0 | 0 | 0 | 2 |
| 9 | 36.39 | 2 | 1 | 1 | 2 | 2 | 2 | 2 | 4 | 100 | 11 | 0 | 1 | 1 | 1 | 0 | 1 | 3 | 0 | 0 | 0 | 1 | 1 | 1 |
| 10 | 38.38 | 2 | 1 | 1 | 0 | 2 | 3 | 1 | 5 | 11 | 100 | 1 | 1 | 0 | 1 | 0 | 1 | 1 | 0 | 1 | 0 | 0 | 0 | 0 |
| 11 | 42.86 | 0 | 1 | 0 | 0 | 0 | 2 | 6 | 3 | 0 | 1 | 100 | 3 | 2 | 0 | 1 | 4 | 3 | 1 | 3 | 2 | 3 | 0 | 3 |
| 12 | 45.16 | 1 | 2 | 1 | 2 | 2 | 3 | 1 | 1 | 1 | 1 | 3 | 100 | 8 | 7 | 0 | 0 | 5 | 3 | 2 | 2 | 0 | 0 | 1 |
| 13 | 46.95 | 0 | 1 | 1 | 1 | 3 | 4 | 0 | 3 | 1 | 0 | 2 | 8 | 100 | 10 | 16 | 2 | 1 | 1 | 2 | 0 | 2 | 3 | 3 |
| 14 | 48.40 | 1 | 2 | 3 | 1 | 6 | 2 | 3 | 0 | 1 | 1 | 0 | 7 | 10 | 100 | 18 | 11 | 2 | 0 | 2 | 3 | 4 | 4 | 3 |
| 15 | 49.10 | 1 | 2 | 3 | 1 | 2 | 1 | 1 | 1 | 0 | 0 | 1 | 0 | 16 | 18 | 100 | 20 | 0 | 1 | 2 | 2 | 3 | 12 | 0 |
| 16 | 51.69 | 0 | 2 | 1 | 1 | 2 | 3 | 1 | 2 | 1 | 1 | 4 | 0 | 2 | 11 | 20 | 100 | 27 | 2 | 0 | 1 | 2 | 8 | 6 |
| 17 | 54.06 | 4 | 1 | 3 | 0 | 1 | 3 | 0 | 1 | 3 | 1 | 3 | 5 | 1 | 2 | 0 | 27 | 100 | 20 | 6 | 4 | 2 | 5 | 3 |
| 18 | 54.45 | 1 | 0 | 0 | 1 | 1 | 2 | 0 | 1 | 0 | 0 | 1 | 3 | 1 | 0 | 1 | 2 | 20 | 100 | 17 | 10 | 0 | 1 | 3 |
| 19 | 55.30 | 0 | 1 | 1 | 0 | 2 | 0 | 2 | 1 | 0 | 1 | 3 | 2 | 2 | 2 | 2 | 0 | 6 | 17 | 100 | 4 | 6 | 1 | 1 |
| 20 | 58.10 | 0 | 3 | 3 | 6 | 2 | 0 | 0 | 0 | 0 | 0 | 2 | 2 | 0 | 3 | 2 | 1 | 4 | 10 | 4 | 100 | 13 | 0 | 2 |
| 21 | 60.15 | 2 | 1 | 3 | 2 | 3 | 2 | 0 | 0 | 1 | 0 | 3 | 0 | 2 | 4 | 3 | 2 | 2 | 0 | 6 | 13 | 100 | 12 | 9 |
| 22 | 61.78 | 0 | 0 | 0 | 0 | 1 | 1 | 1 | 0 | 1 | 0 | 0 | 0 | 3 | 4 | 12 | 8 | 5 | 1 | 1 | 0 | 12 | 100 | 16 |
| 23 | 62.84 | 0 | 1 | 3 | 1 | 0 | 5 | 4 | 2 | 1 | 0 | 3 | 1 | 3 | 3 | 0 | 6 | 3 | 3 | 1 | 2 | 9 | 16 | 100 |

## 5.2.5   SFD 2018: A FRESH LOOK AT EXPERIMENTAL MODAL ANALYSIS

The initial point of departure from pre-2018 SFD practice is estimation of an effective dynamic system over the entire frequency band of interest (rather than selected sub-frequency bands). In order to achieve a satisfactory estimation for the effective dynamic system, the "tolerance" factor ($\varepsilon$) employed in the SVD process is set to a sufficiently low value ($10^{-5}$); in previous sub-frequency band SFD calculations, the SVD "tolerance" factor was set to a value of $10^{-2}$. Computation of effective dynamic system modal parameters, from the first-order system described in Equation (5.17), yields complex modes with eigenvalues having negative and positive imaginary parts. The first level of mode down-selection is to eliminate all modal eigenvalues and eigenvectors that are outside the positive frequency band of interest; for the ISPE modal test, there are 106 complex eigenvalues in the 15–65 Hz frequency band. A vital component of the mode down-selection process is selection of left-hand eigenvectors. $[\Phi_{\eta L}]$ that correspond to their $[\Phi_\eta]$ counterparts; this circumvents issues associated with more involved procedures for computation of a truncated left-hand eigenvector set.

Two computational procedures estimate uncoupled experimental modal FRFs. The first method computes the exact modal solution from the estimated modal parameters of Equation (5.18)d. Specifically,

$$(h_j)_A = (\dot{q}_j)_A = \left( \frac{i 2\pi f}{i 2\pi f - \lambda_j} \right) (\gamma_j). \qquad (5.24)$$

The second method estimates uncoupled experimental modal FRFs from linear combinations of the generalized FRFs (see Equations (5.18)a and (5.18)b) as follows:

$$(h_j)_E = (\dot{q}_j)_E = [\Phi_{\eta L}]_j \cdot (\dot{\eta}). \qquad (5.25)$$

Verification and validation of any candidate estimated experimental mode is now to be judged on the basis of: (a) graphical displays of the modal FRFs and (b) a new modal coherence metric, which is defined as

$$COH_j = \frac{\left|[h_j(f)]_A^* \cdot [h_j(f)]_E\right|^2}{\left|[h_j(f)]_A^* \cdot [h_j(f)]_A\right| \cdot \left|[h_j(f)]_E^* \cdot [h_j(f)]_E\right|}. \tag{5.26}$$

### 5.2.6   ILLUSTRATIVE EXAMPLE: ISPE MODAL TEST

The Integrated Spacecraft Payload Element (ISPE), introduced in Chapter 4, was the subject of modal testing at NASA/MSFC in the fall of 2016. Measured FRF data was quite extensive, and the NASA team had a great deal of difficulty in estimation of modal parameters in the 0–65 Hz frequency band due in part to close modal spacing and significant modal density. These challenges led to development of SFD-2018. For completeness, results of pre-2018 experimental modal analysis methodology and measured FRF reconstruction (similar to the graphic provided in Figure 5.9 for the ISS P5 test article) are illustrated in Figure 5.11.

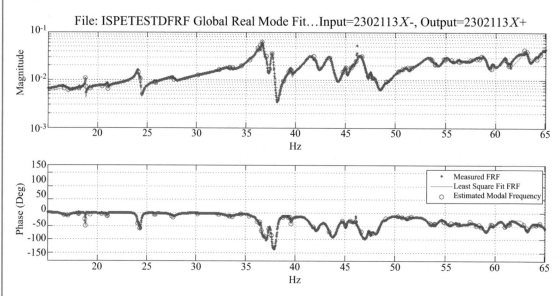

Figure 5.11: Typical ISPE test article MI/MO FRF reconstruction based on pre-2018 SFD methodology.

Establishment of a selected set of experimental modes employing uncoupled SFD-2018 modal FRF estimates, computed using Equations (5.24) (red for reconstructed FRFs) and Equation (5.25) (blue for experimental FRFs) for several candidate modes is illustrated in Figures 5.12–5.14 include FRF magnitude, magnitude and phase, polar, and real and imaginary parts.

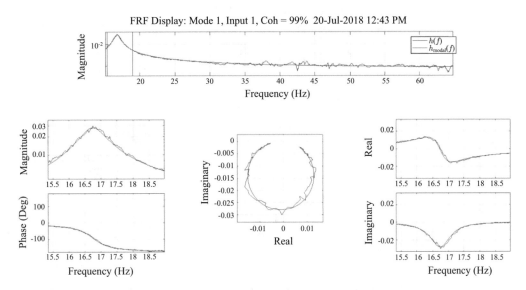

Figure 5.12: EMA for SFD candidate mode 1, excitation 1 (99% coherence).

It is clear in the above three figures that candidate modes 1 and 3 appear valid based on close agreement of the two types of uncoupled modal FRF estimates. In contrast, candidate mode 6 is clearly "spurious."

Table 5.4 summarizes additional information related to the process of selection of 63 acceptable experimental modes from the 106 candidate modes (the first 30 candidate modes and associated modal coherences are illustrated).

Table 5.4 provides a clear demonstration of the utility of the newly introduced EMA metric. In particular, the modal coherence metric, defined in Equation (5.26), for which a value of 85% or greater is assumed to indicate a validly estimated mode.

Returning to the issue of compliance with the NASA-STD-5002 test mode orthogonality criterion, it is quite informative to compare orthogonality matrices computed by: (1) the conventional method based on weighted orthogonality of real modes (in this case the real part of the ISPE test modes) with respect to the TAM mass matrix and (2) exact complex test mode orthogonality assured by left-hand eigenvectors (modes) based on Equation (5.18)b. Results for the 63 ISPE test modes are depicted in Figure 5.15.

While results of orthogonality based on the NASA-STD-5002 criterion indicate good to excellent orthogonality for the first 34 modes, the alternative based on the exact mathematical property of complex SFD-2018 estimated modes opens up the opportunity for automatic satisfaction of the orthogonality criterion without reliance on the (potentially flawed) FEM-based

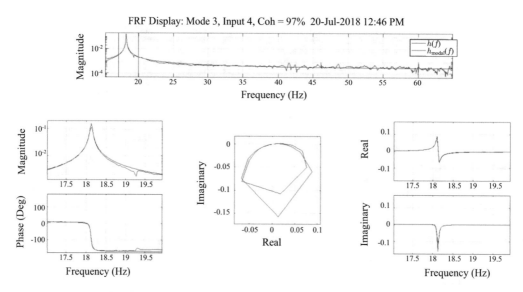

Figure 5.13: EMA for SFD candidate mode 3, excitation 4 (97% coherence).

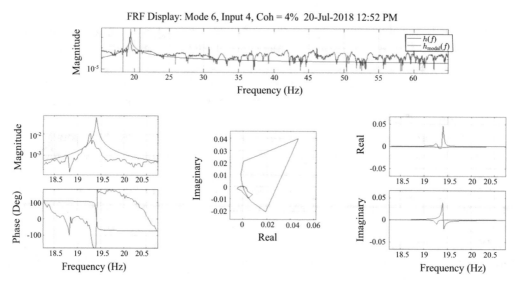

Figure 5.14: EMA for SFD candidate mode 6, excitation 4 (4% coherence).

Table 5.4: EMV modal selection criterion ($COH_j \geq 85\%$) for candidate modes 1–30

| Candidate Mode | Eigenvalue | | Modal Coherences | | | |
|---|---|---|---|---|---|---|
| | Freq (Hz) | Zeta (%) | Excitation 1 | Excitation 2 | Excitation 3 | Excitation 4 |
| 1 | 16.74 | 2.32 | 99.15 | 98.92 | 98.31 | 88.97 |
| 2 | 17.03 | 2.25 | 72.99 | 97.67 | 98.62 | 98.75 |
| 3 | 18.12 | 0.12 | 70.97 | 69.87 | 73.51 | 97.16 |
| 4 | 18.16 | 0.12 | 97.08 | 78.26 | 85.15 | 42.25 |
| 5 | 18.79 | 0.12 | 94.47 | 87.10 | 97.41 | 95.45 |
| 6 | 19.40 | 0.05 | 2.80 | 3.24 | 1.91 | 3.54 |
| 7 | 20.08 | 3.22 | 93.57 | 76.25 | 94.66 | 92.82 |
| 8 | 20.31 | 3.07 | 95.26 | 94.92 | 90.67 | 88.32 |
| 9 | 20.96 | 0.10 | 86.89 | 96.19 | 93.10 | 91.52 |
| 10 | 21.01 | 0.09 | 96.43 | 91.40 | 89.60 | 88.14 |
| 11 | 23.23 | 2.14 | 39.31 | 23.98 | 4.12 | 34.13 |
| 12 | 24.05 | 0.02 | 27.91 | 29.69 | 18.24 | 40.46 |
| 13 | 24.07 | 0.48 | 83.55 | 99.34 | 98.94 | 99.46 |
| 14 | 24.26 | 0.45 | 99.71 | 99.46 | 99.33 | 34.16 |
| 15 | 24.95 | 1.81 | 36.78 | 44.96 | 54.73 | 33.49 |
| 16 | 25.82 | 0.15 | 99.69 | 96.55 | 95.00 | 95.85 |
| 17 | 25.89 | 0.12 | 78.20 | 96.91 | 89.74 | 98.99 |
| 18 | 27.70 | 1.40 | 98.84 | 95.49 | 95.59 | 98.57 |
| 19 | 27.79 | 1.69 | 96.85 | 98.79 | 98.58 | 96.55 |
| 20 | 28.91 | -1.44 | 0.76 | 0.36 | 1.09 | 2.55 |
| 21 | 30.15 | 0.07 | 87.03 | 88.28 | 69.70 | 86.86 |
| 22 | 32.52 | 0.95 | 8.29 | 81.46 | 91.37 | 97.65 |
| 23 | 32.52 | 0.11 | 28.00 | 27.15 | 3.47 | 25.12 |
| 24 | 33.42 | 0.16 | 87.29 | 95.81 | 95.64 | 91.84 |
| 25 | 33.65 | 1.56 | 66.09 | 96.78 | 93.51 | 15.07 |
| 26 | 34.61 | 2.28 | 90.14 | 79.40 | 65.22 | 93.04 |
| 27 | 35.53 | 1.17 | 83.46 | 80.42 | 64.20 | 88.29 |
| 28 | 35.76 | 0.86 | 72.07 | 72.16 | 62.79 | 22.99 |
| 29 | 36.24 | 0.15 | 95.91 | 96.93 | 80.51 | 93.78 |
| 30 | 36.46 | 0.26 | 99.76 | 99.38 | 99.68 | 99.75 |

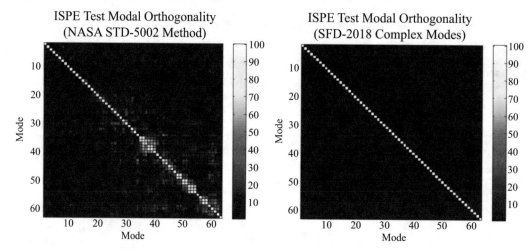

Figure 5.15: ISPE test mode orthogonality estimates.

TAM mass matrix. Further discussion on the merits associated with SFD-2018 is included in Chapter 6.

## 5.2.7  CLOSURE

Experimental modal analysis is a mature discipline in the structural dynamics community, which is as much an "art" as it is a "science." Modern procedures for estimation of modal parameters from measured data are highly automated; however, applications involving complicated structural systems and/or systems with closely spaced, parametrically sensitive modes require the test engineer's experience and judgment ("art") to discern the difference between authentic and spurious ("junk" or "noise") system modes. A prevailing metric for experimental modal data validation is the orthogonality check, which relies on a model-based (TAM) mass matrix. In addition, reconstructive synthesis of measured FRF data is another widely used strategy for experimental mode validation. The present EMA study employs mathematical operations aimed at isolating individual candidate experimental modes without reliance on a TAM mass matrix.

The key to mathematical and visual isolation of individual modes from measured data is the left-hand eigenvector. The most effective approach to determination of left-hand eigenvectors stems from employment of techniques that estimate the measured system's plant or effective dynamic system matrix. Since a complete set of (authentic and "noise") system modes are estimated for the plant, left-hand eigenvectors are determined from the inverse of the complete right-hand eigenvector set.

The following metrics provide a systematic basis for EMA.

1. The estimated SDOF modal FRF, formed by the product of a single estimated left-hand eigenvector and FRF matrix, is plotted in terms of real and imaginary components vs.

frequency, magnitude, and phase components vs. frequency, and polar real vs. imaginary components. Authenticity of an estimated mode is then judged on the basis of quality of the plots.

2. The SDOF modal FRF is also formed from exact mathematical solution of the estimated effective dynamic system. Graphical comparison of this result with the above left-handed product information offers further means of authentic vs. "junk" mode discrimination.

3. Finally, a coherence metric based on comparison of the results of "1" and "2" provides a 0–100% figure of merit for estimated experimental modes.

## 5.3 REFERENCES

[1] R. E. D., Bishop and G. M. L. Gladwell, An investigation into the theory of resonance testing, *Philosophical Transactions, Royal Society of London*, Series A, 225(A-1055):241–280, 1963. DOI: 10.1098/rsta.1963.0004. 85

[2] D. L. Brown and R. J. Allemang, The modern era of experimental modal analysis, *Sound and Vibration Magazine*, January 2007. 85

[3] A. G. Piersol and T. L. Paez, Eds., *Harris' Shock and Vibration Handbook*, 6th ed., McGraw-Hill, 2010. 85, 91

[4] R. Mayes and S. Klenke, The SMAC modal parameter extraction package, *IMAC XVII*, 1999. 92

[5] R. N. Coppolino, A simultaneous frequency domain technique for estimation of modal parameters from measured data, *SAE Paper 811046*, 1981. DOI: 10.4271/811046. 92, 93

[6] R. N. Coppolino and R. C. Stroud, A global technique for estimation of modal parameters from measured data, *SAE Paper 851926*, 1985. DOI: 10.4271/851926. 93

[7] R. N. Coppolino, Efficient and enhanced options for experimental mode identification, *IMAC XXI*, 2003. 92, 93

[8] R. Penrose, A generalized inverse for matrices, *Proc. of the Cambridge Philosophical Society*, 51:406–13, 1955. DOI: 10.1017/s0305004100030401. 92

[9] G. H. Golub and C. Reinsch, Singular value decomposition and least squares solutions, *Numerische Mathematik*, 14:403–420, 1970. DOI: 10.1007/bf02163027. 93

[10] K. Miller, Complex linear least squares, *SIAM Review*, 15(4):706–726, 1973. DOI: 10.1137/1015094. 93

[11] Load analysis of spacecraft and payloads, *NASA-STD-5002*, 1996. 96

[12] Independent structural loads analysis, *U.S. Air Force Space Command, SMC-S-004*, 2008. 96

CHAPTER 6

# Systematic Test Analysis Correlation

## 6.1 PART 1: CONVENTIONAL MASS WEIGHTED CORRELATION OF TEST AND FEM MODES

### 6.1.1 INTRODUCTION

Developments in structural dynamic modeling and modal testing were strongly motivated by aircraft structural failures as early as the first decade of the 20th century. The space race, dating back to the 1950s, and its early failures resulted development of strict government modal test-analysis correlation standards [1, 2] based on mass-weighted orthogonality and cross-orthogonality metrics. The present discussion focuses on (1) review of mass weighted test mode orthogonality and test-analysis cross-orthogonality metrics and (2) introduction of a modal coherence metric. All of the noted metrics are the consequence of a rigorous weighted least-squares formulation [3]. Employment of the conventional mass weighted correlation metrics is demonstrated on modal data associated with two modal test projects.

### 6.1.2 DERIVATION OF MASS WEIGHTED TEST-ANALYSIS CORRELATION METRICS

The relationship between target test modes, $[\Phi_t]$, and their analytical counterparts, $[\Phi_a]$, is described by the transformation

$$[\Phi_t] = [\Phi_a] [COR] + [R], \tag{6.1}$$

where $[COR]$ is the cross-orthogonality matrix and $[R]$ is the residual error matrix. Employing the TAM mass matrix, $[M_{aa}]$, as a weighting matrix, the least squares solution for cross-orthogonality is

$$[COR] = [OR_a]^{-1} \left[ \Phi_a^T M_{aa} \Phi_t \right], \tag{6.2}$$

where the analysis mode orthogonality matrix is

$$[OR_a] = \left[ \Phi_a^T M_{aa} \Phi_a \right] \approx [I], \tag{6.3}$$

and the residual error matrix is orthogonal to the test modes, i.e.,

$$[\Phi_a]^T [M_{aa}][R] \equiv [0].$$

(6.4)

The modal coherence matrix, $[COH]$, is defined based on manipulation of Equation (6.1) (taking into account the results stated in Equations (6.2)–(6.4)) resulting in

$$[COH] = \left([I] - [OR_t]^{-1/2}[R^T MR][OR_t]^{-1/2}\right)$$
$$= [OR_t]^{-1/2}[COR^t OR_a COR][OR_t]^{-1/2},$$

(6.5)

where the test mode orthogonality matrix is

$$[OR_t] = \left[\Phi_t^T M_{aa}\Phi_t\right].$$

(6.6)

The modal coherence matrix, $[COH]$, provides a metric of the ability of mathematical model modes, $[\Phi_a]$, to reconstruct the measured modes, $[\Phi_t]$; in particular, if the diagonal term, $|COH_{nn}|$, associated with measured mode "$n$" is unity, the measured mode is a perfect linear combination of the mathematical model modes, $[\Phi_a]$.

Having already discussed U.S. government criteria for test mode orthogonality, it is informative to note corresponding criteria for test-analysis correlation. NASA-STD-5002 [1] states "Agreement between test and analysis natural frequencies shall, as a goal, be within 5% for the significant modes *(aka target modes)*…Mode shape comparisons shall be required via cross-orthogonality checks using the test modes $[\Phi_t]$, the analytical modes $[\Phi_a]$, and the analytical mass matrix $[M_{aa}]$. The cross-orthogonality matrix is computed as $[\Phi_t]^T[M_{aa}[\Phi_a]$. As a goal, the absolute value of the cross-orthogonality between corresponding test and analytical mode shapes should be greater than 0.9; and all other terms of the matrix should be less than 0.1 for all significant modes." It should be noted that the NASA-STD-5002 [1] definition of cross-orthogonality differs from Equation (6.2), which allows for imperfect TAM mode orthogonality.

It is also informative to note the somewhat stricter U.S. Air Force Space Command criterion [2] for test-analysis correlation, which states, "As a goal, the analytical model frequencies should be within three percent of the measured values, and the cross-orthogonality between the analytical and measured modes, each set normalized to yield a unit generalized mass matrix, should yield values equal to or greater than 0.95 on the diagonal, and equal to or less than 0.10 on the off-diagonal of the cross-orthogonality matrix. Any modeling adjustments/changes made to achieve the above-stated criteria must be consistent with the actual hardware and its drawings."

## 6.1.3    ILLUSTRATIVE EXAMPLE: ISS P5 MODAL TEST

At the time of the ISS P5 modal test, the modal coherence matrix was not employed as part of the modal test-analysis correlation process. It is quite informative nevertheless to review results of cross-orthogonality analysis, which was employed for correlation of (pre-2018 SFD) test modes and FEM analytical modes. As noted in Chapters 3–5, the ISS P5 modal test suffered from challenges associated with localized nonlinearity and non-repeatability of experimental modal data. It is informative to review cross-orthogonality results for the TSS2 data set associated with the ISS P5 pre-test and revised post-test models. A more extensive discussion of the ISS P5 test-analysis reconciliation process (model updating) is provided in Chapter 7. A summary of pre-test ISS P5 modal frequencies, TSS2 data set modal frequencies and the pre-test FEM cross-orthogonality matrix (for the first 10 modes) is provided in Table 6.1.

Table 6.1: ISS P5 data set TSS2 test to pre-test FEM modal frequencies and cross-orthogonality matrix

| | Mode | | | TSS2 Modes | | | | | | | | | |
| | | | | 1 | 2 | 3 | 4 | 5 | 6 | 7 | 8 | 9 | 10 |
| | | Freq (Hz) | | 16.94 | 17.58 | 25.19 | 28.44 | 31.12 | 32.6 | 33.66 | 35.19 | 36.39 | 38.38 |
| Pre-Test FEM | 1 | 18.67 | \|Cross-Orthogonality\| (%) | 83 | 40 | 21 | 1 | 18 | 3 | 4 | 4 | 3 | 0 |
| | 2 | 19.31 | | 47 | 90 | 2 | 2 | 3 | 1 | 1 | 7 | 2 | 0 |
| | 3 | 24.88 | | 23 | 10 | 81 | 1 | 38 | 25 | 1 | 14 | 7 | 1 |
| | 4 | 28.68 | | 2 | 2 | 41 | 63 | 39 | 40 | 10 | 9 | 7 | 2 |
| | 5 | 29.37 | | 4 | 3 | 29 | 36 | 78 | 9 | 19 | 2 | 1 | 6 |
| | 6 | 30.07 | | 0 | 0 | 13 | 55 | 0 | 39 | 49 | 7 | 6 | 2 |
| | 7 | 34.01 | | 4 | 3 | 1 | 18 | 9 | 48 | 53 | 43 | 31 | 29 |
| | 8 | 34.65 | | 1 | 1 | 3 | 4 | 1 | 6 | 4 | 9 | 60 | 84 |
| | 9 | 35.22 | | 1 | 6 | 2 | 10 | 6 | 7 | 8 | 40 | 62 | 41 |
| | 10 | 35.61 | | 7 | 1 | 5 | 4 | 15 | 47 | 51 | 54 | 32 | 17 |

The TSS2 data set pre-test modal frequency correspondence and cross-orthogonality matrix clearly do not satisfy NASA-STD-5002 [1] test-analysis correlation criteria (goals). Model updating operations (to be discussed in Chapter 7) substantially improved the test-analysis correlation situation, as summarized in Table 6.2.

Table 6.2: ISS P5 data set TSS2 test to updated FEM modal frequencies and cross-orthogonality matrix

| | Mode | | | TSS2 Modes | | | | | | | | | |
| | | | | 1 | 2 | 3 | 4 | 5 | 6 | 7 | 8 | 9 | 10 |
| | | Freq (Hz) | | 16.94 | 17.58 | 25.19 | 28.44 | 31.12 | 32.6 | 33.66 | 35.19 | 36.39 | 38.38 |
| Updated FEM | 1 | 16.48 | \|Cross-Orthogonality\| (%) | 98 | 14 | 4 | 0 | 4 | 0 | 1 | 3 | 2 | 1 |
| | 2 | 18.08 | | 8 | 97 | 6 | 0 | 5 | 1 | 2 | 7 | 2 | 1 |
| | 3 | 25.66 | | 4 | 4 | 96 | 17 | 10 | 3 | 4 | 7 | 4 | 1 |
| | 4 | 28.65 | | 1 | 2 | 11 | 92 | 21 | 11 | 1 | 4 | 4 | 3 |
| | 5 | 30.74 | | 2 | 4 | 14 | 18 | 94 | 2 | 7 | 14 | 3 | 4 |
| | 6 | 32.28 | | 2 | 3 | 9 | 12 | 11 | 83 | 22 | 22 | 12 | 7 |
| | 7 | 33.32 | | 2 | 1 | 3 | 0 | 3 | 8 | 82 | 19 | 11 | 8 |
| | 8 | 34.64 | | 1 | 4 | 12 | 9 | 7 | 29 | 17 | 85 | 22 | 3 |
| | 9 | 36.58 | | 1 | 1 | 2 | 2 | 0 | 7 | 7 | 27 | 90 | 14 |
| | 10 | 36.98 | | 0 | 0 | 0 | 3 | 1 | 1 | 0 | 5 | 26 | 97 |

The level of compliance of TSS2 modal test data and the updated P5 FEM model with NASA-STD-5002 test-analysis correlation goals is summarized in Table 6.3.

Table 6.3: ISS P5 data set TSS2 test to updated FEM NASA-STD-5002 compliance status

| Mode | Pre-Test FEM Freq (Hz) | TSS2 Freq (Hz) | Post-Test FEM Freq (Hz) | $\Delta F$(%) | COR (%) | Comments |
|---|---|---|---|---|---|---|
| 1 | 18.67 | 16.94 | 16.48 | -2.72 | 98 | Satisfies NASA-STD-5002 Modal Frequency and On-Diagonal Cross-Orthogonality Goals |
| 2 | 19.31 | 17.58 | 18.08 | 2.84 | 97 | |
| 3 | 24.88 | 25.19 | 25.66 | 1.87 | 96 | |
| 4 | 28.68 | 28.44 | 28.65 | 0.74 | 92 | |
| 5 | 29.37 | 31.12 | 30.74 | -1.22 | 94 | |
| 6 | 30.07 | 32.6 | 32.28 | -0.98 | 83 | Frequency Band Associated with Reduced MI/MO Coherence (strong nonlinearity) |
| 7 | 34.01 | 33.66 | 33.32 | -1.01 | 82 | |
| 8 | 34.65 | 35.19 | 34.64 | -1.56 | 85 | |
| 9 | 35.22 | 36.39 | 36.58 | 0.52 | 90 | Satisfies NASA-STD-5002 Modal Frequency and On-Diagonal Cross-Orthogonality Goals |
| 10 | 35.61 | 38.38 | 36.98 | -3.65 | 97 | |

## 6.1.4    ILLUSTRATIVE EXAMPLE: ISPE MODAL TEST

ISPE modal test data, described in Chapter 5, and ISPE FEM modes are now compared employing the conventional mass weighted correlation procedure. Employing the real part of the ISPE SFD-2018 complex modes and ISPE FEM modes, NASA-STD-5002 test mode orthogonality (Equation (6.3)), cross-orthogonality (Equation (6.2)), and modal coherence (Equation (6.5)) matrices are provided in Figure 6.1.

While the cross-orthogonality matrix indicates poor test-analysis modal correlation for many of the 63 SFD-2018 estimated modes, the modal coherence matrix indicates that approximately 50 of the SFD-2018 modes are linear combinations of the 75 lowest-frequency FEM modes. In Chapter 7, test-analysis reconciliation methodology will demonstrate the potential for drastic improvement of ISPE test-analysis correlation results.

## 6.1.5    CLOSURE

Systematic correlation of FEM based theoretical predictions and experimental modal analysis data, from the viewpoint of a widely accepted U.S. aerospace community practice, relies on weighted least squares based correlation metrics and standardized NASA and USAF criteria. The fundamental assumption inherent in the practice is that test article dynamics may be re-

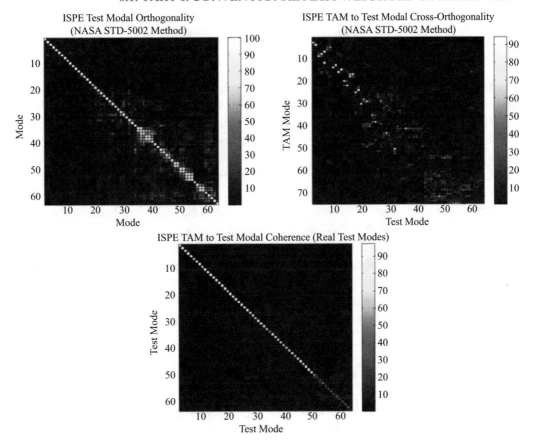

Figure 6.1: ISPE conventional mass weighted test-analysis correlation matrices.

liably approximated in terms of real normal modes, in spite of the facts that (a) a test article may exhibit nonlinear behavior and (b) experimental modes are generally complex. In addition, (c) the correlation metrics are dependent on assumed validity of the TAM mass matrix derived from the theoretical FEM. Due to these inherent assumptions, which at times are questionable, standardized NASA and USAF criteria are specified as goals rather than requirements.

Two test projects serve as illustrative examples for application of the widely accepted test-analysis correlation practice. The ISS P5 modal test, conducted in 2001, demonstrates performance of the weighted least squares correlation metrics in the presence of nonlinear test article behavior. The more recent ISPE modal test, conducted in 2016, demonstrates performance of the weighted least squares correlation metrics in the presence of complex test modes associated with close modal frequency spacing.

## 6.2   PART 2: CORRELATION OF TEST AND FEM MODES USING LEFT-HAND EIGENVECTORS

### 6.2.1   INTRODUCTION

Over many years, members of the experimental modal analysis community have been challenged over the use of NASA and USAF Space Command [1, 2] modal orthogonality and cross-orthogonality criteria for validation of experimental modal vectors and for assessment of test-analysis correlation, respectively. At the heart of the challenge is the role played by the potentially inaccurate TAM mass matrix, which is derived from a mathematical model. Recent work that exploits left-hand eigenvectors, estimated by the SFD technique, provides a promising way out of the TAM mass matrix impasse. Modal orthogonality, defined as the product of left- and right-handed experimental eigenvectors (real or complex) is mathematically an identity matrix. This guarantees that SFD-2018 estimated modes are always perfectly orthogonal. An alternative cross-orthogonality definition, based on weighted complex linear least-squares analysis [3], is evaluated and found to possess the desired property. Employment of (1) the left- and right-handed experimental eigenvector based orthogonality matrix and (2) the weighted complex linear least-squares based cross-orthogonality matrix represents a "game changer" that potentially frees the experimental modal analysis community from the potentially inaccurate TAM mass matrix.

### 6.2.2   MODAL TEST DATA RESULTING FROM ESTIMATED STATE-SPACE MODELS

The SFD-2018 method (discussed in Chapter 5), employs SVD to estimate (real) trial vectors, $[V]$, and generalized acceleration FRFs, $\left[\ddot{\xi}(f)\right]$, associated with measured physical DOF FRFs, $\left[\ddot{U}(f)\right]$, as follows:

$$[\ddot{U}(f)] = [V]\left[\ddot{\xi}_1(f)\right], \qquad [V]^T [V] = [I].$$
(6.7)

Assuming the generalized DOF follow the effective dynamic system form,

$$\left[\ddot{\xi}(f)\right] + [\tilde{B}]\left[\dot{\xi}(f)\right] + [\tilde{K}][\xi(f)] = [\tilde{\Gamma}][F(f)],$$
(6.8)

the effective state-space system modes are computed by solution of the eigenvalue problem,

$$\left\{ \begin{array}{c} \ddot{\xi} \\ \dot{\xi} \end{array} \right\} = \left[ \begin{array}{cc} -\tilde{B} & -\tilde{K} \\ I & 0 \end{array} \right] \left\{ \begin{array}{c} \dot{\xi} \\ \xi \end{array} \right\}, \text{ which is of the general type, } \{\dot{\eta}\} = \left[A_\eta\right]\{\eta\}.$$
(6.9)

Complex eigenvalue analysis of the effective dynamic system produces the generalized complex eigenvectors,

$$\{\eta\} = [\Phi_\eta]\{q\}. \tag{6.10}$$

The "left-hand" generalized eigenvectors are defined as

$$[\Phi_{\eta L}] = [\Phi_\eta]^{-1}. \tag{6.11}$$

Expansion of the generalized eigenvectors using the real trial vectors, $[V]$, defines the test eigenvectors (complex modes) in terms of the physical DOF, specifically,

$$[\Phi]_{TEST} = [V][\Phi_\eta], \qquad [\Phi_L]_{TEST} = [\Phi_{\eta L}][V^T]. \tag{6.12}$$

The state-space eigenvectors resulting from experimental modal analysis using SFD are exactly orthogonal to one another, in particular,

$$[OR]_{TEST} = [\Phi_L]_{TEST}[\Phi]_{TEST} \equiv [I], \tag{6.13}$$

eliminating the need for satisfaction of NASA and/or USAF test mode weighted orthogonality criteria! Moreover, orthogonality of state-space eigenvectors is not dependent on the TAM mass matrix, $[M_{TAM}]$.

## 6.2.3    THEORETICAL SYSTEM MODES IN STATE-SPACE FORM

Modes associated with an undamped or damped theoretical model may be formulated in a manner similar to the considerations employed in the previous discussion. If one substitutes the Guyan [4] (or other suitable) reduction transformation for SVD trial vectors,

$$\{U_{FEM}\} = [\Psi_{TAM}]\{U_{TAM}\}, \tag{6.14}$$

the reduced-order *TAM* modal dynamic equations become

$$[M_{TAM}]\{\ddot{U}_{TAM}\} + [B_{TAM}]\{\dot{U}_{TAM}\} + [K_{TAM}]\{U_{TAM}\} = \{0\}. \tag{6.15}$$

Converting the above system to a state-space description, the following state-space eigenvalue problem is posed:

$$[A_{TAM}][\Phi_{TAM}] = [\Phi_{TAM}][\lambda_{TAM}], \tag{6.16}$$

$$[A_{TAM}] = \begin{bmatrix} -M_{TAM}^{-1}B_{TAM} & -M_{TAM}^{-1}K_{TAM} \\ I & 0 \end{bmatrix}. \tag{6.17}$$

The (generally complex) right- and left-eigenvectors for the state-space TAM automatically satisfy perfect orthogonality in the same manner as the test eigenvectors, specifically,

$$[\Phi_L]_{TAM} = [\Phi]_{TAM}^{-1}, \qquad [\Phi_L]_{TAM}[\Phi]_{TAM} = [I]. \tag{6.18}$$

Results of both experimental and theoretical state-space modal analyses automatically satisfy NASA-STD-5002 and SMC-S-004 orthogonality criteria [1, 2]. Most relevant is the fact that experimental mode orthogonality does not require a TAM mass matrix.

## 6.2.4   CORRELATION OF EXPERIMENTAL AND THEORETICAL MODES (COMPLEX LEAST SQUARES)

Systematic correlation of two sets of state-space, generally complex vectors (with identical DOF designations) is defined in a manner analogous to the case of real vector sets, wherein the transpose operator, is replaced by the complex conjugate transpose [3]. The correlation procedure is formally defined by the following steps:

Step 1: Normalization of complex mode sets

$$[Q_{TEST}] = diagonal\left(\Phi_{TEST}^* \Phi_{TEST}\right)^{-1/2},$$

$$[Q_{TAM}] = diagonal\left(\Phi_{TAM}^* \Phi_{TAM}\right)^{-1/2} \tag{6.19}$$

$$[\Phi_{TEST}]_n = [\Phi_{TEST}] \cdot [Q_{TEST}], \qquad [\Phi_{TAM}]_n = [\Phi_{TAM}] \cdot [Q_{TAM}]. \tag{6.20}$$

Step 2: Pseudo-orthogonalities and pseudo cross-orthogonality

$$[OR_{TEST}] = [\Phi_{TEST}]_n^* [\Phi_{TEST}]_n, \qquad [OR_{TAM}] = [\Phi_{TAM}]_n^* [\Phi_{TAM}]_n \tag{6.21}$$

$$[COR] = [\Phi_{TAM}]_n^* [\Phi_{TEST}]_n. \tag{6.22}$$

Step 3: Modal coherence

$$[C] = [OR_{TAM}]_n^{-1} [COR], \qquad [R] = [\Phi_{TEST}]_n - [\Phi_{TAM}]_n [C] \tag{6.23}$$

$$[COH] = [OR_{TEST}] - [R^* R] = [C]^* [OR_{TAM}] [C] \tag{6.24}$$

Extensive evaluation of the complex least squares process indicates that both the complex cross-orthogonality (Equation (6.22)) and coherence (Equation (6.24)) matrices behave in a manner that parallels the behavior of their real eigenvector counterparts. Therefore, complex modal orthogonality and test-analysis mode correlation may be separated into a "recipe" composed of two distinct sub-tasks.

1. Complex mode orthogonality is perfectly satisfied by left-hand eigenvector counterparts to state-space complex eigenvectors for test and TAM eigenvector sets (Equations (5.19) and (6.13), respectively).

2. Complex mode cross-orthogonality (correlation) and complex modal coherence are defined based on the complex least-squares process (Equations (6.22) and (6.24), respectively).

Due to (a) the presence of many closely spaced shell breathing modes of the test article and (b) non-proportional damping (which is a basic reality of all built-up structural dynamic systems), the test article's modes are generally complex (and are not readily approximated as "real" modes).

## 6.2.5    ILLUSTRATIVE EXAMPLE: ISPE MODAL TEST

The ISPE TAM model is described 75 real, undamped modes in the 0–65 Hz frequency band computed from the reduced TAM model (expressed in state-space form according to Equations (6.15)–(6.18)). ISPE modal test data consists of 63 complex modes in the 0–65 Hz frequency band, estimated by the SFD-2018 algorithm. Following the "recipe" based on Equations (6.13), (6.22), and (6.24), complex mode orthogonality, cross-orthogonality, and modal coherence matrices are summarized in Figure 6.2.

The TAM and test mode orthogonality matrices are perfect identity matrices. The cross-orthogonality matrix appears similar in form to its state-space counterpart (see Figure 6.1), and the modal coherence matrix indicates that the majority of lower frequency ($\sim$50%) of test modes are strong linear combinations of the TAM modes.

Recalling the fact, introduced in Chapter 3, that the orthogonality matrix can be "unpacked" to describe a subject system's modal kinetic energy distribution (Equation (3.3)), a corresponding "unpacking" of the complex state-space mode orthogonality matrix (Equation (6.13)) similarly describes the kinetic energy distribution of the complex modes.

In particular, the modal orthogonality and kinetic energy distribution relationship pair for a complex state-space modal set is defined, in terms of left- and right-hand eigenvectors as,

$$\begin{aligned}
[OR] &= [\Phi_L][\Phi] = [\Phi_{V,L}][\Phi_V] + [\Phi_{U,L}][\Phi_U], \\
[KE_\Phi] &= conj[\Phi_L^*] \otimes [\Phi] = conj[\Phi_{V,L}^*] \otimes [\Phi_V] + conj[\Phi_{U,L}^*] \otimes [\Phi_U].
\end{aligned} \tag{6.25}$$

Note that $[\Phi]^*$ in the present context corresponds to the complex conjugate transpose of the matrix, $[\Phi]$; the same operator designation applies to the left-hand eigenvector matrix. Partitioning of the complex state-space left- and right-hand eigenvectors into "velocity" and "displacement" partitions (designated by the subscripts "$V$" and "$U$", respectively), in order to describe the modal kinetic energy distributions in terms of the measured DOFs, the above definitions of complex state-space eigenvector orthogonality and modal kinetic energy distributions are independent of the TAM mass matrix.

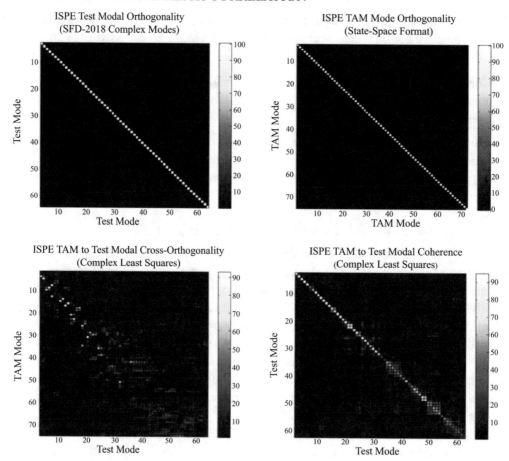

Figure 6.2: Orthogonality, cross-orthogonality coherence of TAM and test complex eigenvectors.

## 6.2.6 ROADMAP FOR A HIGHLY IMPROVED INTEGRATED TEST ANALYSIS PROCESS

Introduction of the complex state-space viewpoint for experimental modal analysis (SFD-2018) and the TAM mass matrix independent metrics described in the present chapter, suggests a potential paradigm shift for the integrated test-analysis process. A roadmap for this envisioned highly improved test analysis process, encompassing experimental modal analysis and systematic test analysis correlation, employing ISPE modal test data and mathematical model data, is outlined in Figures 6.3–6.6.

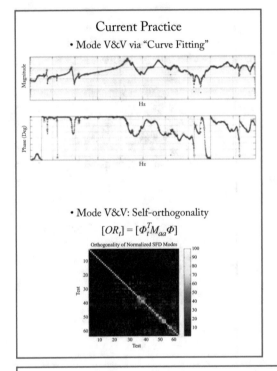

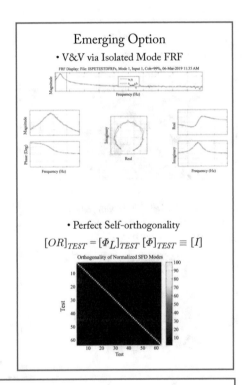

Note: Emerging option is independent of the math model's TAM mass matrix

Figure 6.3: **ISPE** experimental modal analysis and modal orthogonality.

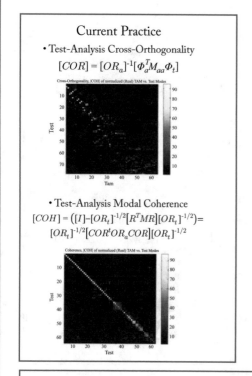

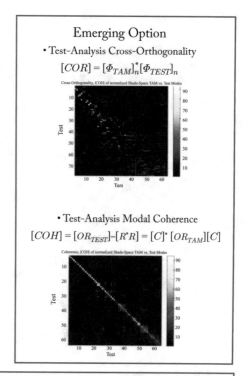

Note: Emerging option is independent of the math model's TAM mass matrix

Figure 6.4: ISPE test-analysis cross-orthogonality and modal coherence.

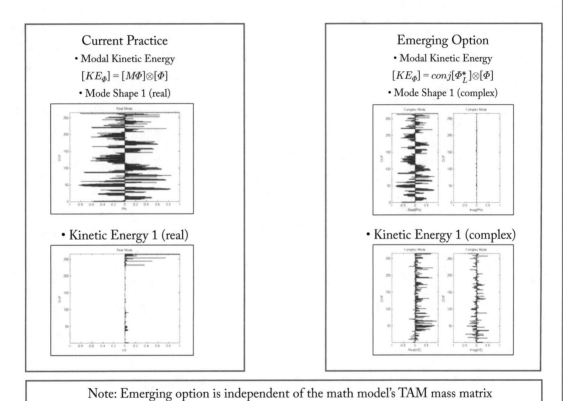

Figure 6.5: **ISPE** test mode 1 mode shape and modal kinetic energy.

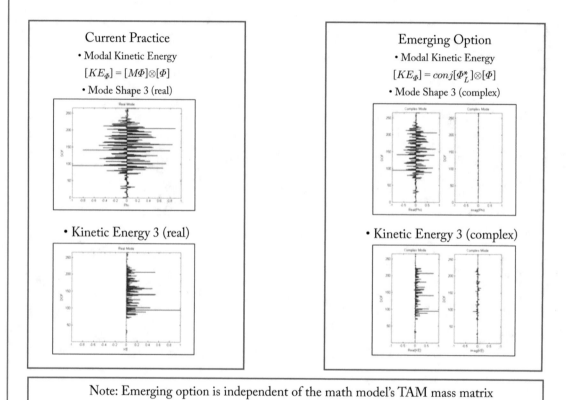

Figure 6.6: ISPE test mode 3 mode shape and modal kinetic energy.

The roadmap for a highly improved integrated test analysis process offers the following innovations.

1. Experimental modal analysis moves from the current, prevailing practice of verification and validation through curve fitting back to past (pre-digital era) practice of single mode isolation made possible by the SFD-2018 technique.

2. Established goals for experimental mode orthogonality (NASA-STD-5002 and USAF Space Command SMC-S-004) are automatically satisfied, without dependence on a possibly flawed and/or inaccurate TAM mass matrix.

3. Test-analysis cross-orthogonality goals (NASA-STD-5002 and USAF Space Command SMC-S-004) are addressed without dependence on a possibly flawed and/or inaccurate TAM mass matrix. Moreover, the modal coherence analysis is defined independent of the TAM mass matrix.

4. While the conventionally used "real" test mode approximation often is similar in content to the corresponding SFD-2018 estimated test mode, modal kinetic energy distributions associated with the "real" and "complex" test modes may differ from one another. This is attributed to the fact that the conventional "real" test mode based modal kinetic energy depends on a possibly flawed and/or inaccurate TAM mass matrix. This is clearly demonstrated by the lack of agreement between conventional and complex modal kinetic energy distributions for ISPE mode 1, and close agreement between conventional and complex modal kinetic energy distributions for ISPE mode 3.

5. The emerging option for a highly improved integrated test analysis process must be reviewed, applied in parallel with established with established practice, and aggressively "poked and prodded" by the technical community before this "paradigm shift" is accepted.

### 6.2.7   CLOSURE

The ISPE modal test experience led to introduction of experimental modal analysis and test-analysis correlation tool enhancements designed to deal with particularly difficult modal testing challenges. Discrimination of valid and spurious experimental modes was addressed in Chapter 5. Test-analysis correlation challenges associated with complex, closely-spaced experimental modes were addressed in the above discussion. Conclusions associated with the newly introduced test-analysis correlation enhancements are as follows.

1. The left-hand, right-hand eigenvector based state-space orthogonality matrix for complex experimental modes automatically satisfies NASA-STD-5002 and USAF SMC-004 requirements. The orthogonality matrix for experimental modes is mathematically perfect by definition, and it is independent of the approximate TAM mass matrix.

2. Employment of the complex least-squares formulation for test-TAM correlation and modal coherence appears to be an appropriate enhancement for incorporation in NASA-STD-5002 and USAF SMC-004 test-analysis correlation standards. It is recognized that the task of modal correlation does not specifically require perfectly orthogonal, real modes.

3. Introduction of the complex least-squares formulation for test-TAM correlation and modal coherence opens the opportunity for inclusion of damping in correlation and update endeavors, e.g., structures with non-negligible modal complexity due to localized, non-proportional damping mechanisms (joints).

The developments influenced by the ISPE test experience offer potential enhancements to U.S. Government modal testing standards that must undergo wider study and critique before establishment of general acceptance in the structural dynamics technical community.

## 6.3   REFERENCES

[1] Load analysis of spacecraft and payloads, *NASA-STD-5002*, 1996. 107, 108, 109, 112, 114

[2] Independent Structural Loads Analysis, U.S. Air Force Space Command, SMC-S-004, 2008. 107, 108, 112, 114

[3] K. Miller, Complex linear least squares, *SIAM Review*, 15(4):706–726, 1973. DOI: 10.1137/1015094. 107, 112, 114

[4] R. Guyan, Reduction of stiffness and mass matrices, *AIAA Journal*, 3, 1965. DOI: 10.2514/3.2874. 113

CHAPTER 7

# Reconciliation of Finite Element Models and Modal Test Data

## 7.1 PART 1: FINITE ELEMENT MODEL MODAL SENSITIVITY

### 7.1.1 INTRODUCTION

Efficient computation of structural dynamic modal frequency and mode shape sensitivities associated with variation of physical stiffness and mass parameters is essential for (1) practical design sensitivity and uncertainty studies and (2) reconciliation of finite element models with modal test data. Sensitivity analysis procedures fall in two distinct categories, namely (a) modal derivatives for small parametric variation and (b) altered system modes associated with "large" parametric variation. The latter category is generally applicable to modal testing, which often requires significant local parameter changes at joints to effect FEM-test reconciliation. However, many investigators and commercial software packages employ estimated modal derivatives in optimization strategies, which address FEM-test reconciliation objectives.

Since the 1960s, methods for computation of modal frequency and mode shape derivatives have evolved. Fox and Kapoor [1] introduced an exact derivative formulation that required knowledge of all modes of the original system; application of the procedure when a truncated set of modes was employed produced compromised derivatives. In response to this difficulty, Nelson [2] derived an exact formulation for computation of mode shape derivatives for truncated mode sets. Efforts to refine and extend application of mode shape derivatives for finite parameter change sensitivity computations have been pursued by many investigators (including the present author). However, the need for modal frequency and mode shape sensitivities that map over very large ranges for multiple parameters suggests application of alternative Ritz [3] strategies.

The Ritz method is one of the most significant developments in analytical mechanics of the past century. This method provides a logical energy formulation for consistent reduction of mass and stiffness matrices employing a set of trial vectors as a reduction transformation. Effectiveness and accuracy of the reduction process depends on selection of an appropriate trial vector set. When a truncated set of baseline system mode shapes is used as the trial vector set (popularly known as Structural Dynamic Modification (SDM)) [4], the Ritz method often produces poor estimates for the altered system. Augmentation of the truncated baseline system mode shapes with appropriately defined additional vectors, however, has been found to produce extremely accurate altered system modal frequencies and mode shapes. Quasi-static residual vectors [5], appended to a truncated set of mode shapes, were found to produce extremely accurate modes for offshore oil platform models subjected to localized alterations [6]. Residual Mode Augmentation (RMA), introduced in 2002 [6] and subsequently refined [7], is a procedure that defines augmented trial vectors, which are appropriate for structures subjected to highly distributed, as well as localized, alterations.

## 7.1.2    SENSITIVITY ANALYSIS STRATEGIES

The present discussion focuses on Ritz procedures that address structural sensitivities due to stiffness and mass alterations described by large (as opposed to small) parametric variations. Therefore, formulations that address computation of eigenvalue and mode shape derivatives are not considered.

The matrix equations describing exact free vibration of baseline and altered structures, respectively, are

$$[K_O][\Phi_O] - [M_O][\Phi_O][\lambda_O] = [0], \tag{7.1}$$

$$[K_O + p \cdot \Delta K][\Phi] - [M_O + p \cdot \Delta M][\Phi][\lambda] = [0]. \tag{7.2}$$

It is implicitly assumed that the stiffness and mass changes scale linearly with respect to the parameter, $p$. Therefore, changes in "beam" depth may not be directly applied, since the axial stiffness (AE) scales linearly with depth and the flexural stiffness (EI) scales as the cube of depth. The appropriate formulation for Equation (7.2) permits linear sensitivity of "AE" and "EI" separately.

The relationship between mode shapes of the baseline and altered structures is expressed as the cross-orthogonality of orthonormal mode shape sets

$$[COR] = [\Phi_O^T][M_O][\Phi], \tag{7.3}$$

where the modal self-orthogonality properties are

$$[OR_O] = [\Phi_O^T][M_O][\Phi_O] = [I_O], \quad [OR] = [\Phi^T][M_O + p\Delta M][\Phi] = [I]. \tag{7.4}$$

The most fundamental Ritz approximation, commonly used in SDM [4] employs a truncated set of low frequency eigenvalues as the reduction transformation described by

$$[\Phi] = [\Phi_{OL}][\varphi]. \tag{7.5}$$

The reduced baseline structure stiffness and mass matrices are, respectively,

$$[k_O] = \left[\Phi_{OL}^T K_O \Phi_{OL}\right] = [\lambda_{OL}], \quad [m_O] = \left[\Phi_{OL}^T M_O \Phi_{OL}\right] = [I_{OL}], \quad (7.6)$$

and the reduced stiffness and mass sensitivity matrices are, respectively,

$$[\Delta k] = \left[\Phi_{OL}^T \Delta K \Phi_{OL}\right], \quad [\Delta m] = \left[\Phi_{OL}^T \Delta M \Phi_{OL}\right]. \quad (7.7)$$

Therefore, the reduced altered structure free vibration equation is

$$[\lambda_{OL} + p \cdot \Delta k] [\varphi] - [I_{OL} + p \cdot \Delta m] [\varphi] [\lambda] = [0]. \quad (7.8)$$

A well-known result of this type of trial vector reduction strategy is that the approximate altered structure eigenvalues are generally higher than results for the exact solution, and the approximate mode shapes do not closely follow the exact shapes when parametric alterations are large.

## 7.1.3   RESIDUAL VECTOR AUGMENTATION FOR LOCAL ALTERATIONS

The static displacements for a baseline structure subjected to unit loads (at physical degrees of freedom where the structure is to be altered) described by the columns of a load array, $[\Gamma]$, are the solutions of

$$[K_O] [U_S] = [\Gamma]. \quad (7.9)$$

A low-frequency modal approximation of static displacements for the above system employs the transformation

$$[U_{SL}] = [\Phi_{OL}] [q_L], \quad (7.10)$$

resulting in the approximate static displacements

$$[q_L] = \left[\lambda_L^{-1}\right] \left[\Phi_{OL}^T\right] [\Gamma], \quad [U_{SL}] = \left[\Phi_{OL} \lambda_L^{-1} \Phi_{OL}^T\right] [\Gamma]. \quad (7.11)$$

The difference between the exact and approximate static solutions defines MacNeal's [5] quasi-static residual vectors,

$$[\Psi_\rho] = [U_S] - [U_{SL}] = \left[K_O^{-1} - \Phi_{OL} \lambda_L^{-1} \Phi_{OL}^T\right] [\Gamma] \equiv \left[\Phi_{OH} \lambda_H^{-1} \Phi_{OH}^T\right] [\Gamma], \quad (7.12)$$

which have been mathematically proven to be the quasi-static displacements associated with all of the high-frequency mode shapes. An orthonormalized set of residual vectors is defined by solution of the residual eigenvalue problem,

$$[k_\rho][\varphi_\rho] - [m_\rho][\varphi_\rho][\lambda_\rho] = [0],$$   (7.13)

where the generalized residual mass and stiffness matrices are

$$[k_\rho] = [\Psi_\rho^T K_O \Psi_\rho], \quad [m_\rho] = [\Psi_\rho^T M_O \Psi_\rho].$$   (7.14)

The augmented trial vector set (replacing the reduction transformation of Equation (7.5))
is

$$[\bar{\Phi}_{OL}] = [\; \Phi_{OL} \quad \Phi_\rho \;],$$   (7.15)

$$[\Phi_\rho] = [\Psi_\rho][\varphi_\rho].$$   (7.16)

When structural alterations are localized, relatively few residual vectors adequately describe the content of changed system mode shapes. The previously described innovation loses its appeal when structural alterations are well-dispersed requiring utilization of many residual vectors.

## 7.1.4   RESIDUAL MODE AUGMENTATION (RMA) FOR DISPERSED ALTERATIONS

Definition of residual vectors associated with dispersed, independent alterations of a baseline structure, described by Equation (7.1), is accomplished by first computing the lowest-frequency mode shapes of the baseline structure (Equation (7.5)) as well as the lowest mode shapes associated with each independent alteration of the structure

$$[K_O + \bar{p}_i \Delta K_i][\Phi_{iL}] - [M_O + \bar{p}_i \Delta M_i][\Phi_{iL}][\lambda_{iL}] = [0] \quad (\text{for } i = 1, \ldots, N).$$   (7.17)

The selected value of each independent scaling parameter is sufficiently large to produce a substantial change in mode shapes (with respect to the baseline structure). An initial set of trial vectors that adequately (and perhaps redundantly) encompass all potential (low frequency) altered system mode shapes is

$$[\Psi] = [\; \Phi_{1L} \quad \Phi_{2L} \quad \ldots \quad \Phi_{NL} \;].$$   (7.18)

This set of trial vectors is expressible as the sum of (a) a linear combination of baseline system mode shapes and (b) trial vectors (that are linearly independent of the baseline system mode shapes)

$$[\Psi] = [\Phi_{OL}][COR] + [\Psi'].$$   (7.19)

The cross-orthogonality coefficient matrix is determined based on the following least-squares solution:

$$\left[\Phi_{OL}^T M_O \Psi\right] = \left[\Phi_{OL}^T M_O \Phi_{OL}\right][COR] + \left[\Phi_{OL}^T M_O \Psi'\right] = [I_{OL}][COR] + [0], \quad (7.20)$$

where

$$[COR] = \left[\Phi_{OL}^T\right][M_O][\Psi], \quad (7.21)$$

$$[\Psi'] = \left[I_{OL} - \Phi_{OL}\Phi_{OL}^T M_O\right][\Psi]. \quad (7.22)$$

The "purified" trial vector set is linearly independent of the baseline system mode shapes in a manner similar to MacNeal's residual vectors, as follows:

$$\left[\Psi'^T M_O \Phi_{OL}\right] = [\Psi^T]\left[I_{OL} - M_O\Phi_{OL}\Phi_{OL}^T\right][M_O\Phi_{OL}]$$

$$= [\Psi^T]\left[M_O\Phi_{OL} - M_O\Phi_{OL}\left(\Phi_{OL}^T M_O\Phi_{OL}\right)\right] \equiv [0] \quad (7.23)$$

$$\left[\Psi'^T K_O \Phi_{OL}\right] = [\Psi^T]\left[I_{OL} - M_O\Phi_{OL}\Phi_{OL}^T\right][K_O\Phi_{OL}]$$

$$= [\Psi^T]\left[K_O\Phi_{OL} - M_O\Phi_{OL}\lambda_{OL}\right] \equiv [0]. \quad (7.24)$$

While the "purified" trial vector set has the above property, it includes an unnecessarily large number of vectors. An appropriate, substantially smaller set of residual vectors is identified by singular value decomposition of the generalized mass matrix,

$$[A] = \left[\Psi'^T M_O \Psi'\right]. \quad (7.25)$$

The singular value decomposition process involves solution of the eigenvalue problem,

$$[A]\left[\varphi_\rho\right] = \left[\varphi_\rho\right]\left[\lambda_\rho\right], \qquad \lambda_{\rho 1} \geq \lambda_{\rho 2} \geq \lambda_{\rho 3} \geq \ldots. \quad (7.26)$$

The cut-off criterion, noted below employed to define suitable reduced trial vector set, is

$$\frac{\lambda_{\rho N}}{\lambda_{\rho 1}} \leq tol = 10^{-N} \qquad \text{(where } N \sim 4\text{–}6 \text{ is usually adequate).} \quad (7.27)$$

The augmented trial vector set (replacing the reduction transformation of Equation (7.5)) is

$$\left[\bar{\Phi}_{OL}\right] = \left[\begin{array}{cc} \Phi_{OL} & \Psi'\varphi_\rho \end{array}\right]. \quad (7.28)$$

The form of the resulting Ritz, multi-parameter sensitivity model (for selected values of the scaling parameters) is

$$\left[k_O + \sum_{i=1}^{N} p_i \left[\Delta k_i\right]\right][\varphi] - \left[m_O + \sum_{i=1}^{N} p_i \left[\Delta m_i\right]\right][\varphi][\lambda] = [0], \qquad (7.29)$$

$$[k_O] = \left[\bar{\Phi}_{OL}^T K_O \bar{\Phi}_{OL}\right], \qquad [m_O] = \left[\bar{\Phi}_{OL}^T M_O \bar{\Phi}_{OL}\right],$$
$$[\Delta k_i] = \left[\bar{\Phi}_{OL}^T \Delta K_i \bar{\Phi}_{OL}\right], \qquad [\Delta m_i] = \left[\bar{\Phi}_{OL}^T \Delta M_i \bar{\Phi}_{OL}\right]. \qquad (7.30)$$

Recovery of mode shapes in terms of physical DOF is accomplished with

$$[\Phi] = \left[\bar{\Phi}_{OL}\right][\varphi]. \qquad (7.31)$$

## 7.1.5   RMA SOLUTION QUALITIES

Since its introduction in 2001, RMA has exhibited the capability to accurately follow modal sensitivity trends over an extremely wide range of parametric variation. The simple cantilevered (planar) beam example, provided in Figure 7.1, demonstrates typical RMA performance ("100%" is baseline). Actual cross-orthogonality checks are also excellent.

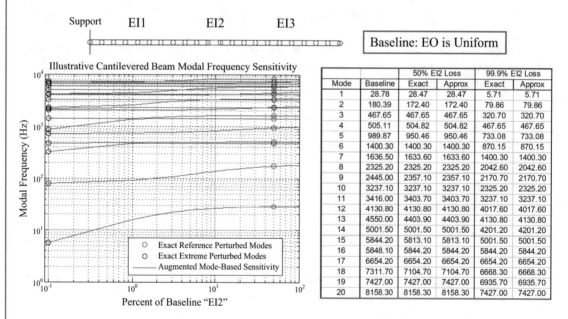

| Mode | Baseline | 50% EI2 Loss | | 99.9% EI2 Loss | |
|---|---|---|---|---|---|
| | | Exact | Approx | Exact | Approx |
| 1 | 28.78 | 28.47 | 28.47 | 5.71 | 5.71 |
| 2 | 180.39 | 172.40 | 172.40 | 79.86 | 79.86 |
| 3 | 467.65 | 467.65 | 467.65 | 320.70 | 320.70 |
| 4 | 505.11 | 504.82 | 504.82 | 467.65 | 467.65 |
| 5 | 989.87 | 950.46 | 950.46 | 733.08 | 733.08 |
| 6 | 1400.30 | 1400.30 | 1400.30 | 870.15 | 870.15 |
| 7 | 1636.50 | 1633.60 | 1633.60 | 1400.30 | 1400.30 |
| 8 | 2325.20 | 2325.20 | 2325.20 | 2042.60 | 2042.60 |
| 9 | 2445.00 | 2357.10 | 2357.10 | 2170.70 | 2170.70 |
| 10 | 3237.10 | 3237.10 | 3237.10 | 2325.20 | 2325.20 |
| 11 | 3416.00 | 3403.70 | 3403.70 | 3237.10 | 3237.10 |
| 12 | 4130.80 | 4130.80 | 4130.80 | 4017.60 | 4017.60 |
| 13 | 4550.00 | 4403.90 | 4403.90 | 4130.80 | 4130.80 |
| 14 | 5001.50 | 5001.50 | 5001.50 | 4201.20 | 4201.20 |
| 15 | 5844.20 | 5813.10 | 5813.10 | 5001.50 | 5001.50 |
| 16 | 5848.10 | 5844.20 | 5844.20 | 5844.20 | 5844.20 |
| 17 | 6654.20 | 6654.20 | 6654.20 | 6654.20 | 6654.20 |
| 18 | 7311.70 | 7104.70 | 7104.70 | 6668.30 | 6668.30 |
| 19 | 7427.00 | 7427.00 | 7427.00 | 6935.70 | 6935.70 |
| 20 | 8158.30 | 8158.30 | 8158.30 | 7427.00 | 7427.00 |

Figure 7.1: RMA sensitivity performance for a cantilevered beam example.

## 7.1.6    ILLUSTRATIVE EXAMPLE: ISPE MODAL TEST

In 2017, an early finite element model of the ISPE test article, provided by Dr. Eric Stewart of NASA/MSFC, was employed for an RMA convergence study. Details of the ISPE model with parametric sensitivity regions are summarized in Figure 7.2.

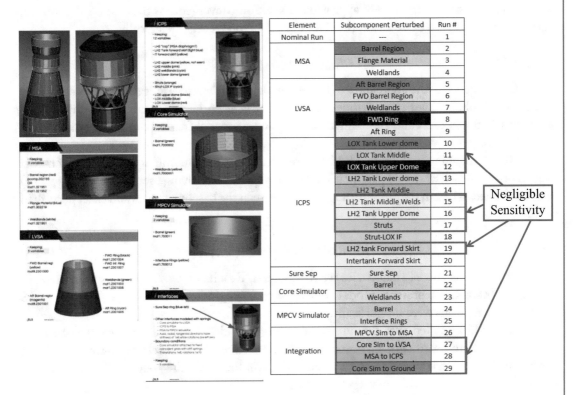

| Element | Subcomponent Perturbed | Run # |
|---|---|---|
| Nominal Run | --- | 1 |
| MSA | Barrel Region | 2 |
| | Flange Material | 3 |
| | Weldlands | 4 |
| LVSA | Aft Barrel Region | 5 |
| | FWD Barrel Region | 6 |
| | Weldlands | 7 |
| | FWD Ring | 8 |
| | Aft Ring | 9 |
| ICPS | LOX Tank Lower dome | 10 |
| | LOX Tank Middle | 11 |
| | LOX Tank Upper Dome | 12 |
| | LH2 Tank Lower dome | 13 |
| | LH2 Tank Middle | 14 |
| | LH2 Tank Middle Welds | 15 |
| | LH2 Tank Upper Dome | 16 |
| | Struts | 17 |
| | Strut-LOX IF | 18 |
| | LH2 tank Forward Skirt | 19 |
| | Intertank Forward Skirt | 20 |
| Sure Sep | Sure Sep | 21 |
| Core Simulator | Barrel | 22 |
| | Weldlands | 23 |
| MPCV Simulator | Barrel | 24 |
| | Interface Rings | 25 |
| Integration | MPCV Sim to MSA | 26 |
| | Core Sim to LVSA | 27 |
| | MSA to ICPS | 28 |
| | Core Sim to Ground | 29 |

Negligible Sensitivity

Figure 7.2: ISPE finite element model and parametric sensitivity regions.

In response to concerns brought up by Dr. Eric Stewart of NASA/MSFC regarding RMA solution convergence, an investigation of the matter was conducted. Specifically, the role of the SVD tolerance parameter (*tol*), defined in Equation (7.26), was evaluated. An objective convergence criterion was developed based on comparison of parametric alterations resulting from the solution of the exact modal equation,

$$\left[ K_O + \sum_i p_i \Delta K_i \right] [\Phi_e] = \left[ M_O + \sum_i p_i \Delta M_i \right] [\Phi_e] [\lambda_e], \qquad (7.32)$$

and the approximate modal equations (see Equations (7.28)–(7.30), developed for a specific value of "*tol*"),

$$\left[ k_O + \sum_i p_i \Delta k_i \right] [\varphi_a] = \left[ m_O + \sum_i p_i \Delta m_i \right] [\varphi_a] [\lambda_a], \quad [\Phi] = [\bar{\Phi}_{OL}] [\varphi]. \qquad (7.33)$$

The metric for evaluation of approximate solution convergence is the cross-orthogonality matrix associated with exact and approximate (RMA) modal sets, specifically,

$$[COR_{ea}] = [\Phi_e^T] [M_O] [\Phi_a]. \qquad (7.34)$$

Convergence of the approximate (RMA) modal set is therefore judged on the basis of how close the (absolute value) cross-orthogonality matrix is to an identity matrix. In addition, the difference between exact and approximate corresponding modal frequencies is also employed as part of convergence evaluation.

Before engaging in the actual RMA convergence study, a preliminary evaluation of modal sensitivities for each of the 28 parametric variations (parameter change set to a value of $p_i = 1$) was conducted, wherein the cross-orthogonality between the baseline modes and exact perturbed modes,

$$[COR_{eo}] = [\Phi_e^T] [M_O] [\Phi_O], \qquad (7.35)$$

and corresponding modal frequencies were evaluated. Results of that exercise, summarized in Table 7.1 (note DF represents modal frequency change), indicate that 13 of the total of 28 parametric variations were insignificant.

It should be noted that the numerical values provided in the above table are the peak frequency and cross-orthogonality alterations associated with the lowest 60 normal modes of each "unit" parametric variation. As a result of this finding, only the "sensitive" 16 parameters were evaluated in the RMA convergence study.

Results associated with the RMA convergence study, which used "unit" parametric variations and values of tolerance (tol) set to 1e-4, 1e-5, and 1e-6, respectively, are summarized in Table 7.2.

It is clear from the above results that $tol = 1e\text{-}6$ produces highly converged RMA modes for the ISPE.

## 7.1.7  CLOSURE

Alteration of a structural dynamic model for the purpose of reconciliation with respect to measured data typically requires moderate to large variation in (stiffness and/or mass) parameters. Moreover, even when small parametric variations of parameters are required, close spacing of system modes produces large variations in modal vectors. Therefore, modal derivatives are not well suited for tracking of parametric sensitivities of structural dynamic modes. A more robust strategy for approximate modal sensitivity analysis employs the Ritz method. Specifically, SDM

Table 7.1: Evaluation of the significance of ISPE model parametric variations

| Case | DF (%) | 100-Cor (%) | Class |
|------|--------|-------------|-------|
| 1.2c.mat | 4 | 31 | |
| 1.3c.mat | 5 | 30 | |
| 1.4c.mat | 9 | 17 | Sensitive |
| 1.5c.mat | 13 | 97 | |
| 1.6c.mat | 13 | 90 | |
| 1.7c.mat | 2 | 35 | |
| 1.8c.mat | 1 | 3 | |
| 1.9c.mat | 1 | 7 | |
| 1.10c.mat | 0 | 0 | Insensitive |
| 1.11c.mat | 1 | 7 | |
| 1.12c.mat | 1 | 10 | |
| 1.13c.mat | 2 | 81 | Sensitive |
| 1.14c.mat | 4 | 22 | |
| 1.15c.mat | 0 | 0 | |
| 1.16c.mat | 0 | 1 | Insensitive |
| 1.17c.mat | 2 | 8 | |
| 1.18c.mat | 23 | 95 | Sensitive |
| 1.19c.mat | 1 | 6 | Insensitive |
| 1.20c.mat | 6 | 93 | |
| 1.21c.mat | 3 | 89 | |
| 1.22c.mat | 2 | 15 | |
| 1.23c.mat | 23 | 100 | Sensitive |
| 1.24c.mat | 16 | 100 | |
| 1.25c.mat | 8 | 2 | |
| 1.26c.mat | 28 | 100 | |
| 1.27c.mat | 0 | 0 | |
| 1.28c.mat | 0 | 0 | |
| 1.29c.mat | 0 | 0 | Insensitive |
| 1.30c.mat | 0 | 0 | |

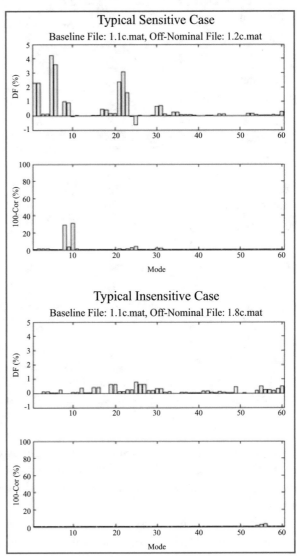

Table 7.2: Summary of RMA convergence study results

| Tolerance | 1.00E-04 | | 1.00E-05 | | 1.00E-06 | |
|---|---|---|---|---|---|---|
| Residuals | 65 | | 160 | | 291 | |
| Case | $|\Delta f|$ (%) | $|\Delta C|$ (%) | $|\Delta f|$ (%) | $|\Delta C|$ (%) | $|\Delta f|$ (%) | $|\Delta C|$ (%) |
| 1.2c.mat | 1.5 | 11 | 1.2 | 10 | 0.1 | 3 |
| 1.3c.mat | 2.3 | 16 | 1.9 | 16 | 0.0 | 1 |
| 1.4c.mat | 3.2 | 33 | 1.6 | 29 | 0.1 | 1 |
| 1.5c.mat | 1.0 | 16 | 0.5 | 1 | 0.2 | 0 |
| 1.6c.mat | 1.3 | 1 | 1.0 | 3 | 0.1 | 0 |
| 1.7c.mat | 0.4 | 8 | 0.4 | 4 | 0.1 | 0 |
| 1.13c.mat | 0.3 | 12 | 0.1 | 2 | 0.0 | 0 |
| 1.14c.mat | 2.5 | 16 | 0.6 | 5 | 0.0 | 0 |
| 1.18c.mat | 0.7 | 19 | 0.1 | 0 | 0.0 | 0 |
| 1.20c.mat | 3.0 | 20 | 1.4 | 7 | 0.1 | 2 |
| 1.21c.mat | 0.7 | 15 | 0.3 | 1 | 0.0 | 1 |
| 1.22c.mat | 0.4 | 7 | 0.2 | 1 | 0.1 | 1 |
| 1.23c.mat | 1.0 | 4 | 0.5 | 3 | 0.1 | 0 |
| 1.24c.mat | 0.1 | 0 | 0.1 | 0 | 0.0 | 0 |
| 1.25c.mat | 1.7 | 99 | 0.6 | 6 | 0.1 | 1 |
| 1.26c.mat | 0.9 | 5 | 0.2 | 2 | 0.0 | 0 |

Notes:
1. $|\Delta f|$ (%) = [approximate−exact frequency]/[exact frequency] (%)
2. $|\Delta C|$ (%) = 100% - [Cross-Othogonality] (%)
3. $|\Delta f|$ and $|\Delta C|$ are the envelopes associated with the lowest 60 system modes

employs a truncated set of baseline system modes as trial vectors to define reduced order mass and stiffness matrices. SDM is a mathematically stable method for approximate parametric sensitivity analysis; however, the baseline system modes are often not adequate for accurate tracking of sensitivities.

RMA is a methodology that defines trial vectors that augment baseline structure modes, resulting in substantially improved the ability of SDM to efficiently track parametric sensitivities of structural modes. RMA employs (a) system modes associated with large reference parametric alterations and (b) SVD to define a reduced set of residual vectors that are rich in dominant geometric changes experienced by the subject structural system. Since its introduction in 2001, RMA has been successfully employed during several modal tests and a variety of mathematical model studies. Most recently, RMA was employed to investigate parametric sensitivities for the ISPE structure. As a result of this work, two enhancements of the method have been introduced.

a. Preliminary screening of candidate parametric sensitivities based on reference frequency shifts and cross-orthogonality metrics was defined to differentiate "sensitive" from "insensitive" cases (eliminating "insensitive" parametric variations from further consideration).

b. A frequency and cross-orthogonality (SVD) convergence metric for determination of an augmented residual trial vector set that satisfies RMA accuracy requirements.

## 7.2    PART 2: TEST-ANALYSIS RECONCILIATION

### 7.2.1    INTRODUCTION

Reconciliation of a test article's finite element model with experimental modal data, if conducted in an objective and systematic manner, requires minimization of a cost function. A variety of modal cost functions are employed by many investigators. The present discussion describes a particular cost function that describes a balanced modal frequency and mode shape "error" relationship. Minimization of the modal cost function's error norm employing gradient based and Monte Carlo strategies are evaluated.

### 7.2.2    MODAL COST FUNCTION ($C_{\phi\lambda}$) AND OPTIMAL TEST-ANALYSIS RECONCILIATION

Consider the standard expression for the undamped structural dynamics eigenvalue problem,

$$[K][\Phi] - [M][\Phi][\lambda] = [0].  \tag{7.36}$$

When modal test data is substituted into the above expression,

$$[K][\Phi_t] - [M][\Phi_t][\lambda_t] = [R],  \tag{7.37}$$

there is a residual error, [R], due to (a) differences between the FEM and modal test data and (2) measurement error.

Pre-multiplication by the FEM mode shapes results in

$$[\Phi^T K \Phi_t] - [\Phi^T M \Phi_t][\lambda_t] = [\Phi^T R].  \tag{7.38}$$

By substituting the transpose of [K][Φ] from Equation (7.36), the above relationship becomes

$$[\lambda][\Phi^T M \Phi_t] - [\Phi^T M \Phi_t][\lambda_t] = [\Phi^T R].  \tag{7.39}$$

Finally, pre-multiplication of the above result by the inverse of the FEM eigenvalues defines the modal cost function [$C_{\Phi\lambda}$] as

$$[C_{\Phi\lambda}] = [\lambda^{-1}\Phi^T R] = [COR] - [\lambda]^{-1}[COR][\lambda_t], \qquad (7.40)$$

where the cross-orthogonality matrix, $[COR]$, associated with test and FEM modes is

$$[COR] = [\Phi^T M \Phi_t]. \qquad (7.41)$$

When there are rigid body and/or zero-frequency mechanism modes present, a shift operator, $\lambda_s$, is employed to avoid numerical problems. The shifted modal cost function is

$$[C_{\Phi\lambda}] = [COR] - [\lambda + \lambda_s]^{-1}[COR][\lambda_t + \lambda_s]. \qquad (7.42)$$

The modal cost function is defined for any corresponding sets of truncated FEM and test-based modal data. It should be noted that when rigid body and/or mechanism modes are not explicitly measured, the set of test modes should be augmented with the corresponding FEM modes.

The FEM may be efficiently altered employing the RMA method discussed in Part 1 of this chapter, for specific values of selected parameters. A reconciled FEM is therefore defined by the combination of selected parameter alterations, which minimize the norm of the modal cost function, $[C_{\Phi\lambda}]$. This may be accomplished by use of a random, Monte Carlo [8] parameter search strategy or the Nelder–Meade Simplex algorithm [9].

### 7.2.3  ILLUSTRATIVE EXAMPLE: RECTANGULAR PLATE SUPPORTED BY CORNER POSTS

The model illustrated in Figure 7.3 is a TAM, defined for a 1386 DOF FEM. Note that the heavy lines illustrate the regions of the model, which are tagged as uncertainties. Two stiffness parameters map the uncertain components: (1) plate edge bending stiffness and (2) elastic support post stiffness. Simulated modal test data (12 modes) was computed based on plate edge bending stiffness increased by 300% and elastic support stiffness reduced by 70%. The baseline FEM of the structure was augmented by sensitivity matrices, which describe the uncertain regions.

In this illustrative example, performances of Monte Carlo and Nelder–Meade Simplex search strategies applied to the modal cost function were explored. Results of the Monte Carlo search analysis are illustrated in Figure 7.4, and results of the Nelder–Meade Simplex search analysis are illustrated in Figure 7.5. Both search strategies reliably estimated the two FEM stiffness parameters.

The bar graph provided in Figure 7.6 illustrates effectiveness of RMA modal sensitivity and the modal cost function (norm) minimization strategy for the present illustrative example.

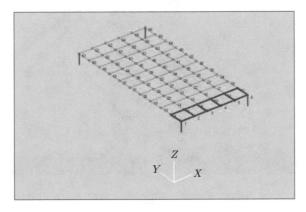

Figure 7.3: Rectangular plate with corner supports.

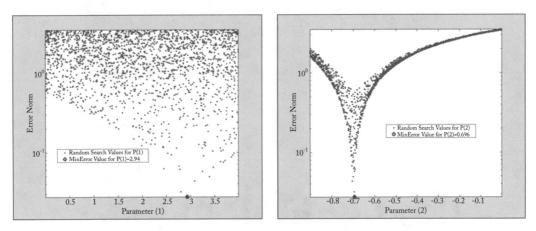

Figure 7.4: Rectangular plate Monte Carlo search analysis patterns.

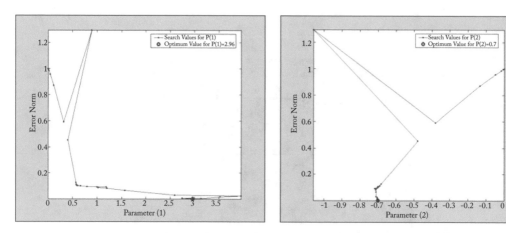

Figure 7.5: Rectangular plate Nelder–Meade simplex search analysis patterns.

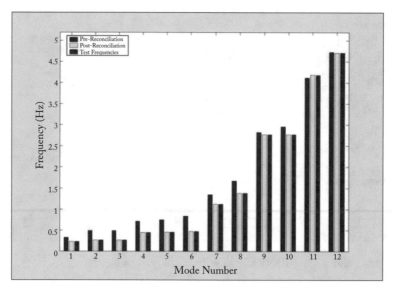

Figure 7.6: Rectangular plate modal cost function (norm) minimization results.

## 7.2.4    ILLUSTRATIVE EXAMPLE: ISS P5 MODAL TEST

The ISS P5 modal test, conducted in 2001, was the first opportunity to employ RMA sensitivity and modal cost function (norm) minimization methodology on a "production" modal test. Before presenting results of this particular investigation, it is important to emphasize valuable "lessons learned" from this first experience.

1. Due to "noise" in the modal test data, gradients associated with the modal cost function, $[C_{\Phi\lambda}]$, were not sufficiently "smooth" for application of the Nelder–Meade Simplex method. Therefore the Monte Carlo search analysis strategy was employed (in this test and all subsequent modal tests since then).

2. Error norm "clouds," which are 2D projections of the multidimensional modal cost function norm and individual parametric variations (all parameters are simultaneously varied in the search process) indicate the relative significance of each parameter (as illustrated in Figure 7.7).

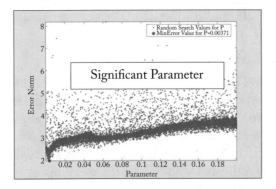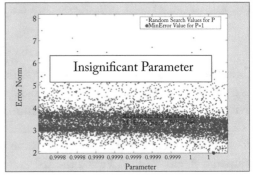

Figure 7.7: ISS P5 typical modal cost function error norms.

More than 25 FEM parameters were initially considered in the modal test-analysis reconciliation exercise. It was initially concluded that nine FEM parameters, when adjusted, effected satisfactory reconciliation for TSS2, TSS4, and TSS17 mode sets. The adjusted parameters are summarized in Table 7.3. Note that only the left forward-mid and mid-aft strut end stiffness ("**LT FM Struts**" and "**LT MA Struts**"), corresponding to joint play were altered from test to test to effect reconciliation.

A summary of TSS2, TSS4, and TSS17 mode set reconciliations indicating compliance with NASA [10] modal frequency and cross-orthogonality goals is provided in Table 7.4.

The previous model reconciliations were judged to be in appropriate compliance with NASA-STD-5002 [10] goals for this challenging modal test (due to significant nonlinear dynamic behavior), by the test team and NASA management. The test-analysis reconciliation exercise was completed within two weeks of the completion of laboratory activities.

Table 7.3: Summary of FEM parameters adjusted for reconciliation

| Parameter | Description | Original Value | TSS02 | | TSS04 | | TSS17 | |
|---|---|---|---|---|---|---|---|---|
| | | | Update Value | % Change | Update Value | % Change | Update Value | % Change |
| 1 | Bot Braces | 3.00E+09 | 0 | -100 | 0 | -100 | 0 | -100 |
| 2 | Lt FM Struts | 3.00E+09 | 1.50E+05 | -99.995 | 6.00E+04 | -99.998 | 3.30E+05 | -99.989 |
| 3 | Lt MA Struts | 3.00E+09 | 2.40E+05 | -99.992 | 1.20E+05 | -99.996 | 3.30E+05 | -99.989 |
| 4 | Grap Lat Spr | Baseline | Baseline | 0 | Baseline | 0 | Baseline | 0 |
| 5 | Mid Long I-1 | 4.56 | 5.077 | 11.3 | 5.08 | 11.3 | 5.08 | 11.3 |
| 6 | Mid Long I-2 | 4.56 | 5.248 | 15.1 | 5.25 | 15.1 | 5.25 | 15.1 |
| 7 | Grap I-1 A | I1(non-uniform) | 1.49XBaseline | 49 | 1.49XBaseline | 49 | 1.49XBaseline | 49 |
| 8 | Grap I-2 A | I2(non-uniform) | 1.46XBaseline | 46 | 1.46XBaseline | 46 | 1.46XBaseline | 46 |
| 9 | Grap I's B | I1&I2(non-uniform) | 1.42XBaseline | 42 | 1.42XBaseline | 42 | 1.42XBaseline | 42 |

Table 7.4: ISS P5 data sets TSS2, TSS4, and TSS17 test to updated FEM NASA-STD-5002 compliance status

| Pre-Test | | TSS02 | | | | TSS04 | | | | TSS17 | | | | Comments |
|---|---|---|---|---|---|---|---|---|---|---|---|---|---|---|
| Mode | Freq (Hz) | Post-Test (Hz) | Test (Hz) | $\Delta F$ (%) | COR (%) | Post-Test (Hz) | Test (Hz) | $\Delta F$ (%) | COR (%) | Post-Test (Hz) | Test (Hz) | $\Delta F$ (%) | COR (%) | |
| 1 | 18.67 | 16.48 | 16.94 | -2.72 | 98 | 15.31 | 15.20 | -0.72 | 98 | 17.31 | 17.37 | 0.35 | 98 | Satisfies NASA STD-5002 Modal Frequency and On-Diagonal Cross-Orthogonality Goals |
| 2 | 19.31 | 18.08 | 17.58 | 2.84 | 97 | 18.02 | 17.92 | -0.55 | 98 | 18.24 | 17.61 | -3.45 | 98 | |
| 3 | 24.88 | 25.66 | 25.19 | 1.87 | 96 | 25.50 | 25.17 | -1.29 | 97 | 25.80 | 25.12 | -2.64 | 94 | |
| 4 | 28.68 | 28.65 | 28.44 | 0.74 | 92 | 28.09 | 28.51 | 1.50 | 92 | 29.16 | 28.30 | -2.95 | 92 | |
| 5 | 29.37 | 30.74 | 31.12 | -1.22 | 94 | 30.56 | 31.32 | 2.49 | 89 | 30.87 | 31.08 | 0.68 | 94 | Frequency Band associated with Reduced MI/MO Coherence (Strong Nonlinearity) |
| 6 | 30.07 | 32.28 | 32.60 | -0.98 | 83 | 31.63 | 32.45 | 2.59 | 84 | 32.65 | 32.66 | 0.03 | 77 | |
| 7 | 34.01 | 33.32 | 33.66 | -1.01 | 82 | 32.93 | 33.49 | 1.70 | 72 | 33.71 | 33.64 | -0.21 | 89 | |
| 8 | 34.65 | 34.64 | 35.19 | -1.56 | 85 | 34.39 | 35.32 | 2.70 | 85 | 34.98 | 35.08 | 0.29 | 87 | |
| 9 | 35.22 | 36.58 | 36.39 | 0.52 | 90 | 36.56 | 36.34 | -0.60 | 92 | 36.60 | 36.32 | -0.77 | 91 | Satisfies NASA STD-5002 Modal Frequency and On-Diagonal Cross-Orthogonality Goals |
| 10 | 35.61 | 36.98 | 38.38 | -3.65 | 97 | 37.00 | 38.58 | 4.27 | 87 | 36.99 | 38.32 | 3.60 | 97 | |

## 7.2.5   ILLUSTRATIVE EXAMPLE: ISPE MODAL TEST

While ISPE FEM-test reconciliation activities have been the responsibility of NASA/MSFC, it is of interest here to engage in an independent hypothetical exercise that employs RMA sensitivity and modal cost function (norm) minimization methodology tools. A hypothetical pair of single parameter mass and stiffness sensitivity matrices, which are "rigged" to effect perfect test-analysis correlation (ISPE modes in the 0–65 Hz band) for a unit value of the sensitivity parameter, $p$, are defined. While the ISPE modes estimated with SFD-2018 are complex, the target test modes for the present exercise are the real parts of those modes.

Migration of ISPE modal frequencies, presented in Figure 7.8, illustrates the capability of RMA to track substantial modal frequency changes associated with large parametric variation (as previously noted in Figure 7.1). Large modal frequency variations with "cross-overs" cannot be adequately tracked through the use of exact and approximate modal derivatives.

Results of Monte Carlo based minimization of the single-parameter-based modal cost function search analysis, and a summary of ISPE original TAM, reconciled TAM, and SFD-2018 test modal frequencies, respectively, are summarized are illustrated in Figure 7.9.

Displays of SFD-2018 test mode orthogonality, TAM-test mode cross-orthogonality, and TAM-test modal coherence, for the original and reconciled TAMs are illustrated in Figure 7.10.

Please note that the above ISPE model correlation and reconciliation exercise represents an idealized hypothetical process. Official ISPE model correlation and reconciliation results, performed by NASA/MSFC, are not presented in this book.

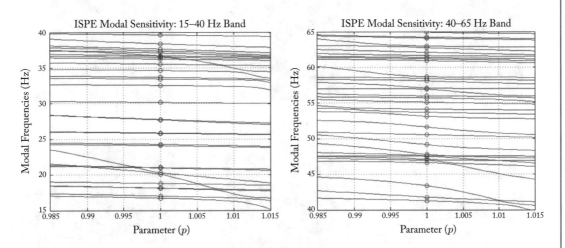

Figure 7.8: Migration of ISPE modal frequencies due to parametric variation.

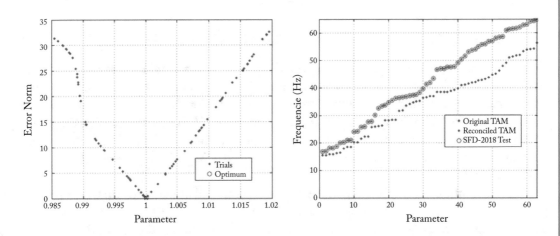

Figure 7.9: ISPE modal cost error norm and ISPE modal frequency summary.

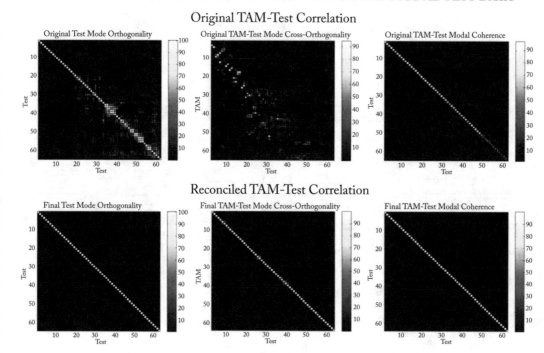

Figure 7.10: Original (pre-test) and reconciled TAM-Test correlation matrices.

## 7.2.6   ILLUSTRATIVE EXAMPLE: WIRE ROPE TEST ARTICLE

While the wire rope test article, discussed in Chapter 4, is not a modal test-correlation and reconciliation application, it provides an example of detailed nonlinear system identification for a dynamic system that possesses strong hysteretic behavior. It is instructive at this point to recall results of preliminary data analysis and nonlinear MI/SO spectral analysis that were presented in Chapter 4. Preliminary data analysis of the estimated isolator deflection and isolator internal force probability density functions is summarized, along with a photograph of the test configuration, in Figure 7.11.

The hypothesized "algebraic" nonlinear system composed of measured time histories (applied force and acceleration response) and synthesized "measured" time histories (cubed displacement and velocity·|velocity|) depicted in Figure 7.12 was subjected to MI/SO analysis.

The cumulative coherence plot, shown in Figure 7.13 indicates that incorporation of the two nonlinear terms produces a nearly unit value cumulative coherence (red curve), while the ordinary coherence associated with a linear model (blue curve) indicates reduced coherence.

While the above results provide clear evidence that the behavior of the wire rope isolators is nonlinear, closer examination of isolator response to swept-sine excitation suggests the wire

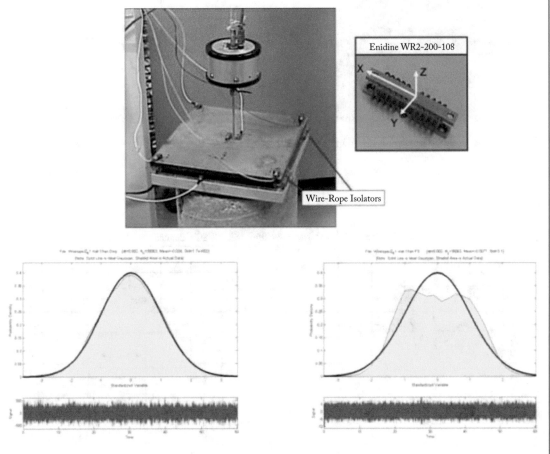

Figure 7.11: Wire rope isolators subjected to broadband random excitation.

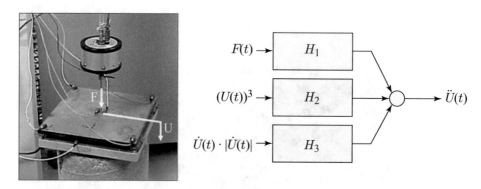

Figure 7.12: Hypothesized MI/SO nonlinear system.

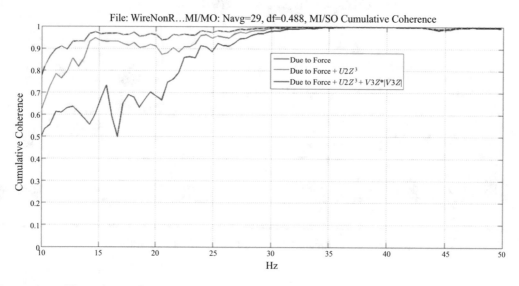

Figure 7.13: Hypothesized nonlinear system cumulative coherence.

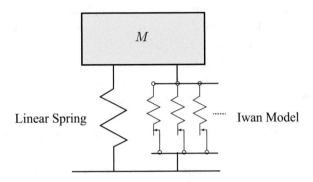

Figure 7.14: **Hypothesized nonlinear wire rope model.**

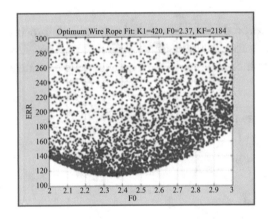

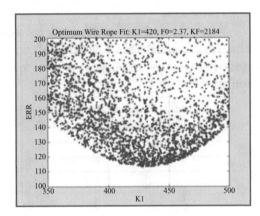

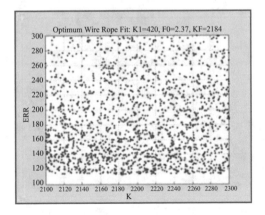

Figure 7.15: **Wire rope cost function error norm projections.**

rope isolator exhibits "hysteretic" nonlinear behavior. The hypothesized wire rope model consists of the linear spring and Iwan friction slip [11] element depicted in Figure 7.14.

Unknown parameters for the model consist of (a) linear spring stiffness, $K1$, (b) total Iwan stiffness, $KF$, and (c) total Iwan critical slip force, $F0$; the Iwan parameters are uniformly distributed among $n = 5$ sub-elements.

Estimation of the three unknown parameters was accomplished by minimization of the error norm defined by the square of the absolute difference between measured (swept-sine) and model force time histories. A Monte Carlo search strategy was employed to determine optimum values for the three unknown parameters. Error norm projections for the three parameters are illustrated in Figure 7.15.

Effectiveness of the wire rope nonlinear system identification process is illustrated in comparison of the measured and fitted model load-deflection plots provided in Figure 7.16. The dashed and solid black lines indicate stiffness asymptotes associated with non-slip and fully slipped behaviors.

While the majority of this book has been devoted to the integrated test analysis process for linear structural dynamic systems, this final illustrative example offers a glimpse of promising future developments that may expand the process to include systems with quite general nonlinear features.

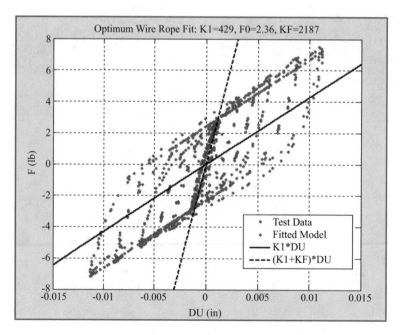

Figure 7.16: Results of the wire rope hysteretic nonlinear model system identification.

## 7.2.7 CLOSURE

Efficient and accurate reconciliation of finite element models with experimental modal data is first of all dependent on high quality data sets associated with models and measured data sets, assured by adherence to all steps of the integrated test analysis process presented in this book. Careful formulation of predictive finite element models of a subject system's anticipated operational and test article configurations, taking full advantage of modern CAE/FEM resources, is essential. Specifically, finite element models must fully conform with design drawings and supporting data, paying special attention to joint and interface details (to facilitate appropriate parametric sensitivity analyses). Objective assessment of unexpected test results followed by repeated diagnostic tests and evaluations, as illustrated in the ISS P5 modal test program, are essential in many situations. Success of the integrated test analysis process is highly dependent upon willingness of the management and technical staff members to ponder unexpected and inconvenient experimental data, rather than force/adjust results in conformity with preconceived notions.

On the assumption that the above prerequisite guidelines are followed, the two essential tools for modal test analysis reconciliation are (1) accurate, efficient finite element sensitivity models and (2) robust cost function definition and optimizations. This book features an effective RMA procedure that appears to eliminate inaccuracies and ambiguities inherent in techniques based on modal derivatives and truncated modal data sets, aka SDM. Indeed, RMA, which is gaining a high level of acceptance in the U.S. aerospace technical community, provides a valued enhancement for SDM. A variety of projects over the past 18 years, beginning with the ISS P5 modal test, have pointed to advantages of Monte Carlo over Nelder–Meade optimization due to the presence of non-smooth gradients in test-analysis cost functions owing to "noise." While this author has employed one specific cost function definition for test-analysis reconciliation, the reader should be aware of other cost function definitions and optimization strategies used by experienced technical organizations.

While the primary emphasis in this book centers on modal test applications, the wire rope test and system identification example, points to strategies aimed at objective characterization of nonlinear dynamic phenomena.

Finally, it is recommended that serious attention should be given to the "roadmap for a highly improved integrated test analysis process" presented in Chapter 6 of this book. The "roadmap" offers a way around (1) dependence on a pre-test model based TAM mass matrix (assuring perfectly orthogonal experimental modes) and (2) limitations associated with the unrealistic "real-modes" approximation for closely spaced modes and the "many modes" problem. In order to properly exploit the latest SFD-2018 methodology, test analysis reconciliation strategies must be defined and evaluated to address the "reality" of complex experimental modes.

## 7.3    REFERENCES

[1] R. I. Fox and M. P. Kapoor, Rates of change of eigenvalues and eigenvectors, *AIAA Journal*, 6, 1968. DOI: 10.2514/3.5008. 123

[2] R. B. Nelson, Simplified calculation of eigenvector derivatives, *AIAA Journal*, 14, 1976. 123

[3] W. Ritz, Über eine neue methode zur Lösung gewisser variationsprobleme der mathematischen physik, *Journal für die reine und angewandte Mathematik*, 135, 1908. DOI: 10.1515/crll.1909.135.1. 123

[4] Twenty years of structural dynamic modification—a review, *IMAC 20*, 2002. 124

[5] R. MacNeal, A hybrid method for component mode synthesis, *Computers and Structures*, 1, 1971. DOI: 10.1016/0045-7949(71)90031-9. 124, 125

[6] R. N. Coppolino, Structural mode sensitivity to local modification, *SAE Paper 811044*, 1981. DOI: 10.4271/811044. 124

[7] R. N. Coppolino, Methodologies for verification and validation of space launch system (SLS) structural dynamic models, *NASA CR-2018–219800*, 1, 2018. 124

[8] N. Metropolis and S. Ulam, The Monte Carlo method, *Journal of American Statistical Association*, 44(247):335–341, 1949. DOI: 10.2307/2280232. 134

[9] J. Nelder and R. Mead, A simplex method for function minimization, *Computer Journal*, 7(4):308–313, 1965. DOI: 10.1093/comjnl/7.4.308. 134

[10] Load analyses of spacecraft and payloads, *NASA-STD-5002*, 1996. 137

[11] W. Iwan, On a class of models for the yielding behavior of continuous composite systems, *Journal of Applied Mechanics*, 89:612–617, 1967. 144

# Author's Biography

## ROBERT N. COPPOLINO

**Robert N. Coppolino** received his formal education in New York City having graduated from Stuyvesant High School and earned a B.S. Aerospace Engineering, and an M.S. and Ph.D. in Applied Mechanics at the Polytechnic Institute of Brooklyn. As a structural dynamics engineer at Grumman Aerospace Corporation (1967–1975), he supported the Lunar Module, Skylab, and Space Shuttle programs, and contributed to development of techniques for launch vehicle pogo (instability) suppression and propellant tank hydroelastic modeling and test-analysis correlation. From 1975–1983, Dr. Coppolino was at The Aerospace Corporation, ultimately holding the post of manager of the Engineering Dynamics Section. While at Aerospace, he supported NASA's Space Shuttle Pogo Integration Panel, developed the Simultaneous Frequency Domain (SFD) method for experimental modal analysis, and defined innovative residual vector techniques the were successfully employed in Space Shuttle nonlinear payload interface flight loads predictions and detection of structural damage on offshore jacket platforms. From 1983–1987, he served as manager of the Advanced Methods and Development branch at the MacNeal–Schwendler Corporation (MSC). During his stay at MSC he led a consortium composed of MSC, the J. S. Bendat Company, and Synergistic Technology Incorporated, which developed technologies that were ultimately called the Integrated Test Analysis Process (ITAP). From 1987 to the present, Dr. Coppolino has held various senior posts at Measurement Analysis Corporation, currently serving as Chief Technology Officer for Torrance Operations. During this period he collaborated closely with the late Julius S. Bendat on development of nonlinear spectral techniques for identification of nonlinear systems. In addition, Dr. Coppolino continued development of ITAP components including efficient procedures for modal test-analysis reconciliation. The full suite of ITAP capabilities experienced its first end-to-end, "prime-time" application on the International Space Station P5 Short Spacer modal test, conducted at NASA/MSFC in 2001.

In 2016, Dr. Coppolino was invited to join the NASA Engineering Safety Center, Loads and Dynamics Technical Discipline Team (LDTDT) as an industry member. Since that time, he has provided independent review of NASA's SLS and Orion programs, and as part of this work developed procedures aimed at dealing with the "many modes" problem. A highlight of these activities is development of SFD-2018, which appears to provide the impetus for a Roadmap for a Highly Improved Modal Test Process.

Robert Coppolino is married to Catherine (Stafford) Coppolino, a person of outstanding courage and love for others. Together they have five children, Michael, Melisssa, Kenneth, Peter, and James, who have blessed them with seven wonderful grandchildren, Hannah, Hailey, Jakob, Robert, Ryder, Drew, and Tyler.

Printed in the United States
by Baker & Taylor Publisher Services